Science Foundations

Physics Plus

Bryan Milner

CAMBRIDGE

PUBLISHED BY THE PRESS SYNDICATE OF THE UNIVERSITY OF CAMBRIDGE
The Pitt Building, Trumpington Street, Cambridge, United Kingdom

CAMBRIDGE UNIVERSITY PRESS
The Edinburgh Building, Cambridge CB2 2RU, UK
40 West 20th Street, New York, NY 10011-4211, USA
477 Williamstown Road, Port Melbourne, VIC 3207, Australia
Ruiz de Alarcón 13, 28014 Madrid, Spain
Dock House, The Waterfront, Cape Town 8001, South Africa

http://www.cambridge.org

First published 2002

Printed in the United Kingdom at the University Press, Cambridge

Typeface Stone Informal   *System* QuarkXPress®

A catalogue record for this book is available from the British Library

ISBN 0 521 89237 6 paperback

Designed, edited and produced by Gecko Limited, Cambridge

Illustrations by John Batten, Helen Humphreys, Geoff Ward and Pete Welford

ACKNOWLEDGEMENTS
H. Rogers/Art Directors & Trip 9, 11, 12tr, 13r, 34tr, 34br, 42, 43, 52; A. Lambert/Art Directors & Trip 10, 12bl, 13l, 13cr, 28, 30; Darwin Dale/Science Photo Library 12tl; Aidan Gill 12cl; Amy Trustram/Science Photo Library 13cl; Neil Thompson 14, 20; Mark Campbell/Photofusion 34bl; D. Rule/Art Directors & Trip 57; Popperfoto/Reuters 67l, 73b; Rosenquist/Earth Pictures/Camera Press, London 67tr; Corbis 67br, 73t; David Parker/Science Photo Library 72.

The publisher has made every effort to trace copyright holders, but if they have inadvertently overlooked any they will be pleased to make the necessary arrangements at the earliest opportunity.

Contents

■ **How to use this book**   4

■ **Physics in action**

Electronic control systems

1   Controlling currents in circuits _____ 6

2   Controlling electrical appliances
     automatically: the thermostat _____ 8

3   Electronic control systems 1:
     the basic idea _____ 10

4   Electronic control systems 2:
     extending the range _____ 12

5   Looking at processors _____ 14

6   Using logic gates in electronic
     control systems _____ 16

7   Another way of looking at
     processor inputs and outputs _____ 18

8   How to switch on large currents ____ 20

9   How to connect input sensors 1:
     switches _____ 22

10   How to split voltages:
     the potential divider _____ 23

11   How to connect input sensors 2:
     LDRs and thermistors _____ 24

12   Making a moisture sensor _____ 26

13   How to adjust input sensor circuits __ 27

H1   Putting it all together _____ 28

14   What's the delay? _____ 30

H2   More about time delay circuits _____ 32

H3   Are electronic systems a good
     thing? _____ 34

Optical devices

15   Two types of lens _____ 36

16   Two types of image _____ 38

H4   More about real images _____ 40

H5   Bigger – but not real! _____ 42

■ **Forces and motion**

1   Turning forces _____ 44

2   When the swinging stops _____ 46

H1   Balanced turning forces _____ 48

H2   Why do things fall over? _____ 50

3   How to accelerate without going
     faster _____ 52

4   Circular motion 1: in the solar
     system and beyond _____ 54

5   Circular motion 2: inside atoms ____ 55

6   Circular motion 3: spin driers _____ 56

7   Circular motion 4: cornering on
     a cycle _____ 57

8   Momentum _____ 58

H3   Changing momentum _____ 59

H4   Explosions _____ 60

H5   Collisions _____ 62

9   The Earth's structure _____ 64

(I+E) 10   Movements that make mountains __ 66

11   How does the Earth's crust move? ___ 68

12   What keeps the Earth's crust
     moving? _____ 70

(I+E) 13   Earthquakes _____ 72

(I+E) 14   Changing ideas about the Earth ____ 74

H6   More about tectonic plates _____ 76

■ **What you need to remember:
completed passages**   78

■ **Glossary/index**   82

How to use this book

A note about the teaching sequence of modules in AQA Separate Science Physics

It is assumed in this book:

■ that the *Physics in action* module will be taught <u>after</u> the modules *Energy* and *Electricity*;

■ that the *Forces and motion* module will be taught <u>after</u> the modules *Waves and radiation* and *Forces*.

■ An introduction for students and their teachers

This book comprises two chapters:

■ *Physics in action*

and

■ *Forces and motion*.

These chapters match the two physics modules, additional to those in Double Award Science, needed for AQA GCSE Separate Science Physics.

The material in each chapter is split into separate topics, numbered in order.

Each topic normally takes up a double-page spread, though some topics comprise just a single page.

There are two types of topic in each chapter:

■ topics comprising scientific ideas that <u>all</u> Separate Science Physics students are expected to know (in addition to those in Double Award Science), whether they are entered for the Foundation Tier or the Higher Tier of Separate Science Physics tests and examinations;

■ topics comprising the additional scientific ideas that <u>only</u> candidates entered for the Higher Tier Separate Science Physics tests and examinations need to know.

Because most Separate Science students are entered for the Higher Tier of GCSE, the topics required by the Higher Tier only are interleaved with the topics required by all students.

This provides better continuity. These Higher Tier topics are numbered H1, H2, H3 etc. both in the *Contents* and at the top of the page.

Occasionally, there is not enough Higher Tier material for a whole topic. You will then find a box, like the one shown below, in a topic that is otherwise required by all students.

> This material is always inside a brown border.

For Higher Tier students only

In some topics, there are ideas from your previous studies that you will be building on but which there isn't the space to explain fully again.

You will find these ideas briefly summarised like this:

> This material is always inside a purple border.

Ideas you need from *Electricity*

> You will be given the name of the module you have previously studied.

■ Science that all Separate Science Physics students need to know

Physics in action

Most of the material in the book is of this type. It does not have any special border or heading.

1

Controlling currents in circuits

The answers to these questions are provided in the *Supplementary Materials*.

Each time you are introduced to a new idea you will be asked a question.
This is so you can make sure that you really understand the ideas.

At the end of each topic you will find a section like this.

What you need to remember [Copy and complete using the key words]

You should keep your answers to these sections in a separate place. They contain all the ideas you are expected to remember and understand in tests and examinations. So they are very useful for revision.

It is very important that these summaries are correct, so you should always check your summaries against those provided on pages 78–81 of this book.

■ Science that only Higher Tier students need to know

Physics in action

There will be a heading like this.

H1 For Higher Tier students only

Putting it all together

This material is always inside a brown border.

You will find questions in the text. Your answers to these questions will provide you with a summary of the additional ideas that you are expected to remember and understand for Higher Tier tests and examinations. You should keep them with your 'What you need to remember' summaries so you can use them for revision.

Because the answers to these questions are a summary of what is on the extension pages, no further answers are provided.

At the end of each topic you will find a section like this.

Using your knowledge

The questions in these sections are like many of the questions you will meet in Higher Tier tests and examinations. You have to use ideas from the topic to explain something new. You are not expected to remember the answers to these questions.

Answers to these questions are provided in the *Supplementary Materials*.

■ A note about practical work

Practical work, where you observe things and find out things for yourself, is an important part of Science. You will often see things in this book which you have yourself seen or done, but detailed instructions for practical work are not included. These will be provided separately by your teacher.

The *Supplementary Materials* contain many suggestions for practical activities.

■ A note about *Ideas and evidence*

All GCSE Science specifications must now assess candidates' understanding of what the National Curriculum calls *Ideas and evidence*. Those parts of this book which deal with this aspect of Science are indicated, on the *Contents* page, like this:

I+E H3 Are electronic systems a good thing? ____ 34

1

Controlling currents in circuits

■ Switching currents on and off

The most important thing you need to be able to do with an electrical circuit is to switch the current on and off.

For an electric current to flow, there must be a **complete** circuit of **conductors** between the two sides of the power supply (or cell, or battery).

If you put a switch into a circuit, you can then:

■ switch the current on by completing the circuit

■ switch the current off by **breaking** the circuit.

1 Copy and complete the table.

Switch	Symbol	Circuit (complete/broken)	Lamp (on/off)
open			
closed			

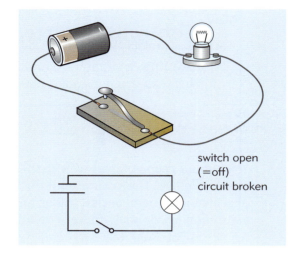

switch open
(=off)
circuit broken

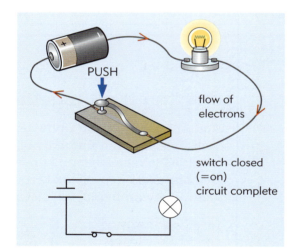

PUSH

flow of electrons

switch closed
(=on)
circuit complete

■ Increasing and reducing currents

Sometimes we don't just want to be able to switch a current on and off.

We also want to be able to increase the current or to reduce it.

2 Describe, as fully as you can, <u>two</u> examples of where it is useful to be able to change the size of a current.

To dim the light when watching TV, you need to reduce the current through the lamp.

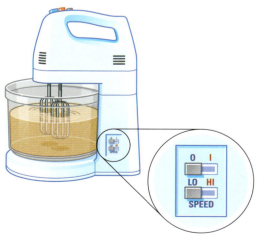

When you change the speed of a mixer, you are changing the current through its motor.

How to increase and reduce currents

One way of changing the size of a current is to change the **resistance** of the circuit (see Box).

3 Copy and complete the sentence.

   To reduce the current through a circuit you must _____ the resistance of the circuit.

Two ways of changing resistance

The diagrams show two different ways of changing the resistance of a circuit.

4 For each circuit, write down:

   (a) what you have to do to change the resistance

   (b) what happens to the other component in the circuit when you <u>reduce</u> the resistance.

Problems with using resistance to control current

Reducing the current through a circuit by increasing the resistance of the circuit is not very efficient and may sometimes be unsafe (see Box).

5 Explain why using resistors is not a very satisfactory way of controlling currents.

There is a different way of using fixed and variable resistors to control circuits so that only very small currents flow through them (see pages 23–27). The resistors do not then become hot.

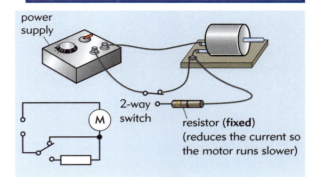

power supply

2-way switch

resistor (**fixed**) (reduces the current so the motor runs slower)

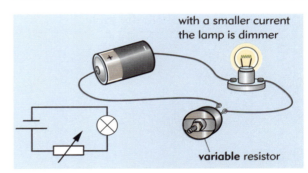

with a smaller current the lamp is dimmer

variable resistor

Too hot to handle

When a current flows through a resistance, energy is transferred as heat (thermal energy).

This means that the resistors used to control currents can become hot. It also means less energy is transferred by the circuit in the way that you want. In other words, the circuit is less efficient.

What you need to remember [Copy and complete using the **key words**]

Controlling currents in circuits

For an electric current to flow from a battery or power supply you need a _____ circuit of _____ .

You can switch off a current by _____ the circuit.

You can reduce the current flowing through a circuit by increasing the _____ of the circuit.

You can do this using a _____ resistor   [draw the symbol]

or a _____ resistor.   [draw the symbol]

2 Controlling electrical appliances automatically: the thermostat

Controlling appliances by hand

With many electrical appliances, we like to be able to switch them on and off, or to control them, by hand. We call this **manual** control.

Nowadays, the manual control of many appliances can be done <u>remotely</u>.

1 (a) Write down the names of <u>four</u> appliances that people usually control manually.

(b) Which of these appliances can be controlled remotely?

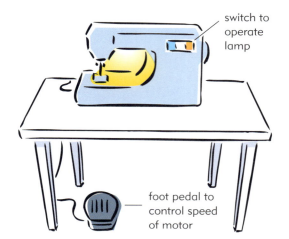

Controlling appliances automatically

Sometimes, however, manual control is very inconvenient.

For example, we normally want the temperature of a room to stay the same for long periods of time.

But we don't want to have to watch the temperature all the time and keep switching the heating on and off.

We want a system that does this <u>automatically</u>.

2 Write down the names of <u>three</u> other appliances that need to have automatic temperature control.

A device that automatically keeps things at a steady temperature is called a **thermostat**.

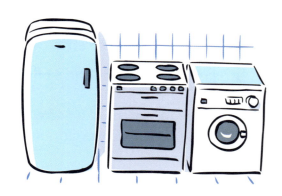

3 Why is *thermostat* an appropriate name for this device?

Looking at a thermostat

Thermostats have been used for many years in heating appliances and heating systems.

The diagram shows a thermostat from an electric iron.

4 Make a copy of the flow diagram.
Then write the following sentences in the correct boxes to explain how the thermostat works.

The circuit is broken

The heating element makes the iron hotter

The iron cools down

The bi-metal strip bends away from the contact

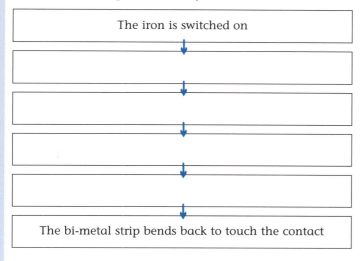

5 How is the thermostat adjusted so that the iron is

(a) slightly hotter?

(b) slightly cooler?

The thermostat shown on this page works because the bi-metal strip bends when the temperature changes. We say that the thermostat is a <u>mechanical</u> device.

Nowadays more and more automatic control systems are **electronic**.

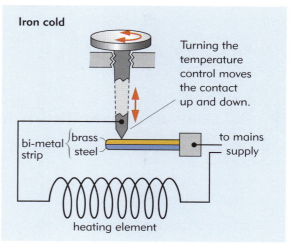

Iron cold

Turning the temperature control moves the contact up and down.

bi-metal strip { brass steel

to mains supply

heating element

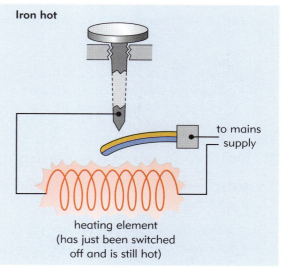

Iron hot

to mains supply

heating element (has just been switched off and is still hot)

Brass expands more than steel. So the bi-metal strip bends away from the contact and breaks the circuit.

What you need to remember [Copy and complete using the **key words**]

Controlling electrical appliances automatically: the thermostat

For some purposes, hand-operated (_____) controls are very convenient.
Sometimes, however, we prefer automatic control systems, for example temperature controlled by a _____.
Nowadays automatic control systems are often _____.

3 Electronic control systems 1: the basic idea

Nowadays, automatic control systems are used for many different things. These control systems are usually electronic.

Ideas you need from *Electricity*

Thermistor LDR

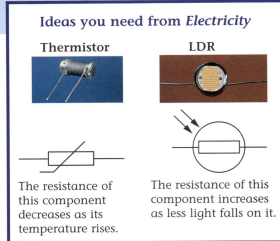

The resistance of this component decreases as its temperature rises.

The resistance of this component increases as less light falls on it.

■ A light that automatically switches on in the dark

The diagram shows an electronic system that automatically switches on a light when the surroundings are dark.

This type of diagram is called a block diagram.

The diagram shows how the output from one stage of the system provides the input to the next stage of the system.

1 Make a copy of the block diagram.
   Change the captions to describe what happens when the surroundings are light.

Block diagram of an electronic system
(for switching a light on and off automatically)

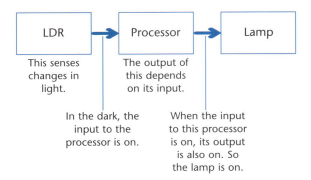

LDR → Processor → Lamp

This senses changes in light.

The output of this depends on its input.

In the dark, the input to the processor is on.

When the input to this processor is on, its output is also on. So the lamp is on.

■ The three stages of any electronic system

Electronic systems can be made to control many different things.

But all electronic systems consist of the same three stages. These are shown in the general block diagram.

2 Look at the automatic lighting system shown on this page.

   (a) What is the output device?

   (b) What component is used as an input sensor?

   (c) Explain why this component is suitable.

   (d) Copy and complete the table to show how the processor responds.

Input	Output
on	
off	

General block diagram
(for any electronic system)

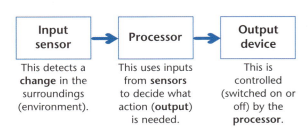

Input sensor → Processor → Output device

This detects a **change** in the surroundings (environment).

This uses inputs from **sensors** to decide what action (**output**) is needed.

This is controlled (switched on or off) by the **processor**.

An electronic thermostat

The diagram shows an electronically controlled heating system.

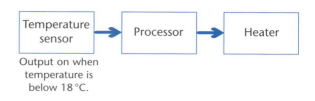

Output on when temperature is below 18 °C.

3 (a) What is the output device?

   (b) What component could you use as an input sensor?

4 Copy and complete the sentences.

When the temperature of the surroundings falls below 18 °C, the input to the processor from the input sensor goes to _____.

The processor then switches the heater _____.

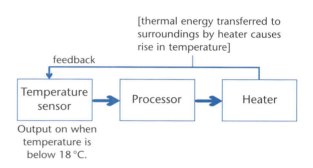

[thermal energy transferred to surroundings by heater causes rise in temperature]

feedback

Output on when temperature is below 18 °C.

The electronic thermostat works because of <u>feedback</u> from the heater to the temperature sensor.

5 When the heater is switched on, the temperature of the surroundings eventually rises to above 18 °C.

Write <u>two</u> sentences, similar to those in question 4, to describe what then happens.

6 In the automatic light switch shown on the opposite page, you must <u>prevent</u> feedback from the output device to the sensor.

   (a) <u>How</u> can you prevent feedback?

   (b) <u>Why</u> must you prevent feedback?

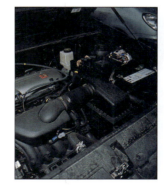

Electronic control systems are being used to do more and more jobs in cars. They are not only used to control the performance of the engine and the brakes but also to switch on the lights and windscreen wipers automatically.

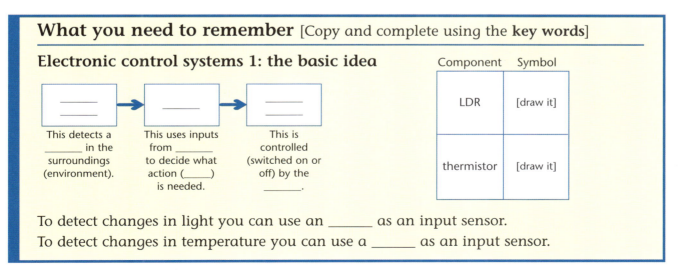

What you need to remember [Copy and complete using the **key words**]

Electronic control systems 1: the basic idea

____ → ____ → ____

This detects a _____ in the surroundings (environment).

This uses inputs from _____ to decide what action (_____) is needed.

This is controlled (switched on or off) by the _____.

Component	Symbol
LDR	[draw it]
thermistor	[draw it]

To detect changes in light you can use an _____ as an input sensor.

To detect changes in temperature you can use a _____ as an input sensor.

11

4

Electronic control systems 2: extending the range

We want to be able to make a wide variety of electronic control systems, to do many different jobs. We need:

- a range of input sensors that respond to many different aspects of the environment

- a range of output devices that act on the environment in different ways.

Some of the input sensors and output devices that we can use are shown on this page and the next.

LED

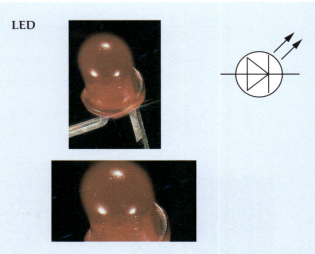

LEDs need only a very small current, but they only emit a small amount of light.

Motor

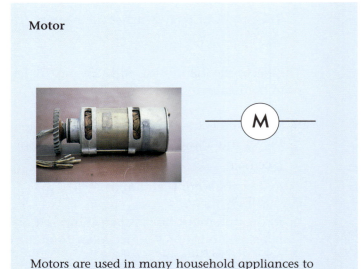

Motors are used in many household appliances to transfer energy as movement (kinetic energy).

Buzzer

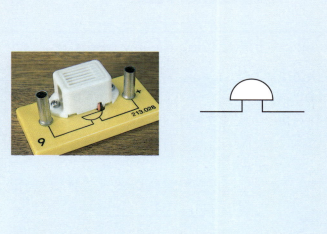

Buzzers transfer energy to the surroundings as sound.

Push switch

PUSH ➡

For this switch to stay on, you have to keep applying pressure.

LDR

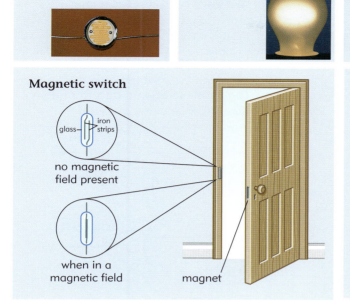

Lamp

Thermistor

Heater

Magnetic switch

glass — iron strips

no magnetic field present

when in a magnetic field

magnet

Tilt switch

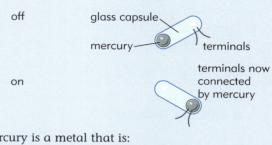

off

glass capsule

mercury — terminals

on

terminals now connected by mercury

Mercury is a metal that is:
■ liquid at normal room temperatures
■ an electrical conductor.

1 Write down <u>one</u> use for LEDs.

2 Name <u>five</u> electrical appliances that use electric motors.

3 What <u>two</u> properties of the metal mercury make it very suitable for use in a tilt switch?

4 Explain, as fully as you can, how a magnetic switch can be used to detect whether a door is open or closed.

5 (a) Select from the devices shown on these pages, a suitable input sensor and output device to use in:

   (i) a burglar alarm

   (ii) a visual indicator on a freezer that shows it is at the correct temperature

   (iii) an automatic system which pumps up liquid to a header tank to keep the correct level of liquid.

  (b) Draw block diagrams for each of the above systems.

What you need to remember

Electronic control systems 2: extending the range

Make a copy of the following tables, then complete them by including all the input sensors and output devices shown on these two pages. Also copy and complete the sentence below the tables.

Input sensor	What it responds to

Output device	Transfers energy to surroundings as…

The symbol for an LED is: [draw it]

5

Looking at processors

So far we have looked at the various different input sensors and output devices.
Now it's time to have a look at the devices that are used as processors.

Processors are usually made from devices called **logic gates**.

Three types of logic gate

The diagrams show the three types of logic gate that you need to know about.

1 Copy and complete the table.

Name of gate	How many inputs?	How many outputs?

Logic gate circuits are built on silicon chips.
The photograph shows a chip which contains several logic gates.

2 Write down <u>three</u> reasons why logic gates are often used.

The three types of logic gate can also be combined in various ways. This makes them very versatile.

AND gates

The information in the Box describes what an AND gate does.

3 (a) What input(s) must there be to an AND gate for its output to be on?

(b) Why is this gate called an AND gate?

4 Which of the AND gates P, Q, R and S have an output that is on?

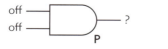 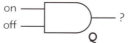  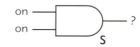

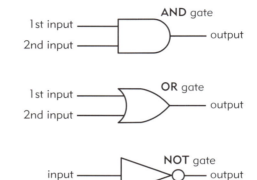

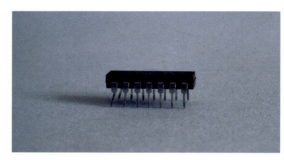

Logic gates are reliable and cheap to make. This small LC chip contains several logic gates.

For the output of an AND gate to be on:

the 1st input must be on

and

the 2nd input must be on.

OR gates

The information in the Box describes what an OR gate does.

5 (a) What input(s) must there be to an OR gate for its output to be on?

(b) Why is this gate called an OR gate?

6 Which of the OR gates W, X, Y and Z have an output that is on?

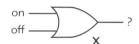

NOT gates

The information in the Box describes what a NOT gate does.

7 (a) What input must there be to a NOT gate for its output to be on?

(b) Why is this gate called a NOT gate?

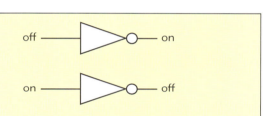

For the output of a NOT gate to be on the input must **not** be on.

What you need to remember [Copy and complete using the **key words**]

Looking at processors

The processors in electronic control systems are usually made from _____ _____.

Name of gate	Symbol	For the output of the gate to be on ...
_____		the 1st input must be on _____ the 2nd input must be on [or <u>both</u> inputs must be on]
_____		the input must _____ be on
_____		the 1st input must be on _____ the 2nd input must be on

Using logic gates in electronic control systems

In an electronic control system, you must have the right logic gate for the job you want it to do.

REMEMBER

For the output of an AND gate to be on:
the 1st input must be on **and** the 2nd input must be on.

For the output of an OR gate to be on:
the 1st input must be on **or** the 2nd input must be on [or <u>both</u> inputs must be on].

For the output of a NOT gate to be on:
the input must **not** be on.

■ Example I

The diagram shows an electronic alarm system.

1 (a) What type of logic gate is used?

   (b) What input(s) does this gate need to make its output on?

   (c) What must happen so that the gate receives these inputs?

   (d) What is the system designed to do?

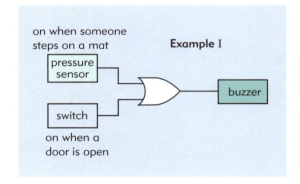

on when someone steps on a mat

Example I

pressure sensor

switch

on when a door is open

buzzer

Older burglar alarm systems for houses often have sensors on <u>all</u> doors and windows to detect when they are opened or disturbed in any way.

Nowadays, burglar alarms usually use sensors that detect the infrared radiation emitted from the warm body of any intruder inside a house.

■ Example II

The diagram shows an electronic system that controls a lamp inside a car.

2 (a) What type of logic gate is used?

   (b) What input(s) does this gate need to make its output on?

   (c) What must happen so that the gate receives these inputs?

   (d) What is the system designed to do?

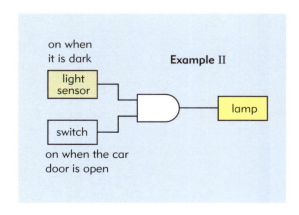

on when it is dark

Example II

light sensor

switch

on when the car door is open

lamp

Example III

The top diagram shows an electronic system that controls the temperature inside a room (or, in other words, an electronic thermostat).

3 (a) What inputs does the gate need so that it turns on the heater?

   (b) How does the gate receive these inputs when there is only <u>one</u> input sensor?

4 (a) Which of the systems A–D would also work as an electronic thermostat for the heating system?

   (b) For each of the systems that would <u>not</u> work, explain why it wouldn't work.

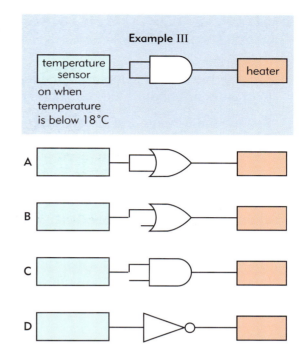

Example IV

The diagram shows an electronic system that could be used to control the temperature inside a freezer.

5 (a) What type of logic gate is used?

   (b) What input does this gate need to make its output on?

   (c) What must happen so that the gate receives this input?

   (d) What is the system designed to do?

A note about large currents

Electric heaters require a large current.

Electronic systems can normally supply only a small current.

How this problem can be solved is explained on pages 20–21.

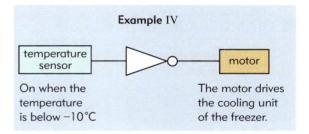

Example IV

temperature sensor

On when the temperature is below −10°C

motor

The motor drives the cooling unit of the freezer.

What you need to remember

Using logic gates in electronic control systems

You must be able, when you are presented with a block diagram of a simple electronic system, to describe:

■ what each part of the system does

■ what the whole system does

just as you have done on these pages.

7 Another way of looking at processor inputs and outputs

So far in this book, the inputs to and the outputs from logic gates have been called 'on' and 'off'.

Because these inputs and outputs can only have the two values, 'on' or 'off', they can be represented by the numbers 1 (= on) and 0 (= off).

The numbers 1 and 0 are called binary digits.

1 Suggest why 0 and 1 are called binary digits. [Hint: think about a <u>bi</u>cycle.]

The way that the output of a logic gate depends on the input(s) can be shown using 0s and 1s in a table (see Box). A table of this kind is called a <u>truth table</u>.

> ## A note about 'high' and 'low'
>
> 'on' (or 1) is sometimes called 'high'
>
> 'off' (or 0) is sometimes called 'low'
>
> The reason for this is explained on page 20.

> ## REMEMBER
>
> For the output of an AND gate to be on:
> the 1st input must be on
> **and**
> the 2nd input must be on.
>
> For the output of an OR gate to be on:
> the 1st input must be on
> **or**
> the 2nd input must be on
> [or <u>both</u> inputs must be on].
>
> For the output of a NOT gate to be on
> the input must **not** be on.

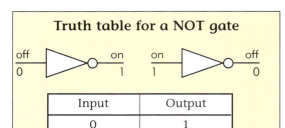

Truth table for a NOT gate

Input	Output
0	1
1	0

▪ Truth table for an AND gate

The diagrams show the output from an AND gate for all the different inputs it can have.

2 Copy and complete the truth table for the AND gate.

1st input	2nd input	Output
0	0	
1	0	
0	1	
1	1	

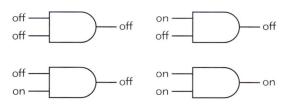

▪ Truth table for an OR gate

The diagrams show the output from an OR gate for all the different inputs it can have.

3 Make a truth table for the OR gate.

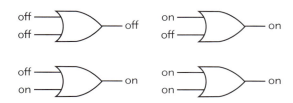

Combinations of gates

You can use truth tables to work out what happens when two, or more, gates are connected together.

In Example I, the NOT gate simply reverses the output of the AND gate.

You can show this by using the truth table for the AND gate on page 18 and adding a fourth column for the final output.

4 Make a truth table for Example I.

Example II is a bit trickier.

The Box explains how you can work out the truth table for this combination of gates.

5 Copy and complete the following.

D in Example II is:
■ the output of logic gate _____
■ the input to logic gate _____.

6 Gates X and Y in Example II are then swapped around.

Make a truth table for this combination of gates.

Representing problems as truth tables

Suppose that you want a bleeper to sound if you open a car door with the headlights still switched on.

The Box shows how you can write down this problem in the form of a truth table.

You can then see what logic gate you need to use.

7 (a) What logic gate has the same truth table as the one that shows the car bleeper problem?

(b) What is the point of this electronic system?

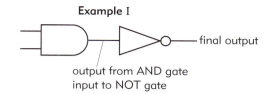

Example I

output from AND gate
input to NOT gate

final output

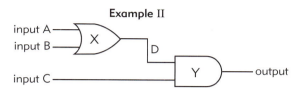

Example II

input A
input B
X
D
input C
Y
output

First write down all <u>eight</u> possible combinations of inputs A, B and C. Then use A and B to work out D. Finally use C and D to find the output.

A	B	C	D	Output
0	0	0	0	0
1	0	0	1	0
0	1	0	1	0
1	1	0	1	0
0	0	1	0	0
1	0	1	1	1
0	1	1	1	1
1	1	1	1	1

The problem in words...			...and as a truth table		
Headlight	Door	Bleeper	Headlight	Door	Bleeper
off	shut	off	0	0	0
on	shut	off	1	0	0
off	open	off	0	1	0
on	open	on	1	1	1

Bleep, bleep, bleep...
(You've left the lights on)

What you need to remember

Another way of looking at processor inputs and outputs

You need to be able to make, and to interpret, truth tables in the ways that you have on these pages.

How to switch on large currents

So far in this book, we have considered electronic control systems only in the form of block diagrams.

We must now look at some of the details of the actual circuits that are used in electronic control systems.

One important consideration is that most electronic circuits can only handle a small **current**.

But many output devices require quite large currents.

1 Give an example of an output device that needs a large current.

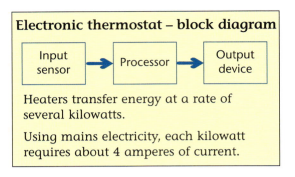

Electronic thermostat – block diagram

Input sensor → Processor → Output device

Heaters transfer energy at a rate of several kilowatts.

Using mains electricity, each kilowatt requires about 4 amperes of current.

■ Solving the problem

To solve the problem of providing a large current, a buffer is needed between the electronic control system and the output device.

The diagram shows what can be used to provide such a buffer.

2 What device is used to provide a buffer?

relay

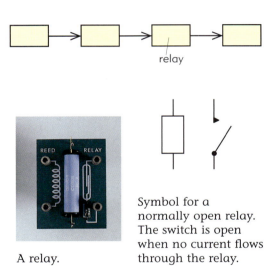

A relay.

Symbol for a normally open relay. The switch is open when no current flows through the relay.

The diagram shows why the **relay** acts as a buffer.

3 Copy and complete the sentences.

The electronic control system sends a _____ current through the relay.

This current operates a _____ inside the relay.

When the switch is closed, a _____ current flows through the output device.

The relay keeps the large current to the output device completely separate from the electronic control system.

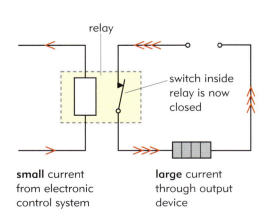

relay

switch inside relay is now closed

small current from electronic control system

large current through output device

How a relay works

The diagrams show how one type of relay works.

4 Write down the following sentences in the correct order to explain how the relay works.

- The contact completes the circuit to the output device.

- The output from the processor goes to 'on'.

- The armature is attracted by the electromagnet.

- A large current flows through the output device.

- The coil and its iron core become magnetised.

- A small current flows through the coil of the relay.

5 The relay shown in the diagram is called a <u>normally open</u> relay. Suggest why.

Another use for a relay

Electronic control circuits operate from a **low** voltage d.c. supply. But we often want to use them to control appliances that work from the **mains** supply. This creates a problem even when the mains operated output device needs only a small current.

6 Explain why there is a problem.

Using a relay as a buffer solves the problem by keeping the low voltage d.c. supply and the 230 volt mains supply completely separate.

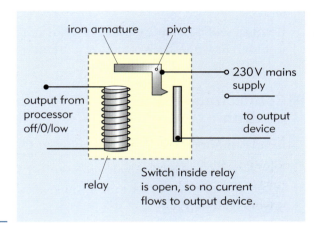

Switch inside relay is open, so no current flows to output device.

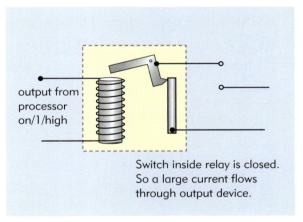

Switch inside relay is closed. So a large current flows through output device.

What you need to remember [Copy and complete using the **key words**]

How to switch on large currents

Electronic control circuits can usually supply only a small _____ .
To control output devices that need a large current, a _____ is used as a buffer.
This device works because it uses a _____ current to switch on a _____ current.
Relays can also be used to allow _____ voltage electronic circuits to control appliances that work from the 230 volt _____ supply.

The symbol for a relay is: [draw the symbol for a normally open relay with its switch open]

9

How to connect input sensors 1: switches

There are many different types of switch.
What must happen to a switch to make it go 'on' or
'off' is different in different cases.

> **1** What must happen to each of the switches shown in
> the diagrams to make them go 'on'?
> (You may need to look back at page 13.)

Because switches can respond to what happens to them
in different ways, they make very useful input sensors.

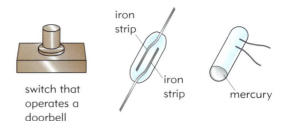

switch that
operates a
doorbell

iron
strip

iron
strip

mercury

Three types of switch.

■ Connecting switches into electronic control circuits

In an ordinary electric circuit you use a switch to turn
a <u>current</u> on or off.

When you use a switch as an input sensor, however,
it works differently.

It allows you to connect the **positive** side of a low
voltage supply to an **input** of a logic gate.

When the switch is on, this voltage is big enough for
the input to the gate to be on (or 1, or **high**).

> **2** The inputs to logic gates are sometimes called 'high'
> and 'low' rather than 'on' and 'off'. Explain why.
>
> **3** Look at Diagram I.
> Then copy and complete the sentences.
>
> In Diagram I, the switch is _____ so the input to
> the OR gate is _____.
>
> This means that the output from the gate is _____
> so the LED is _____
>
> **4** Write <u>two</u> similar sentences for Diagram II.

> **REMEMBER**
>
> The output of an OR gate is on if either,
> or both, of the inputs is on.

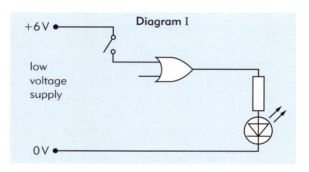

+6V

low
voltage
supply

Diagram I

0V

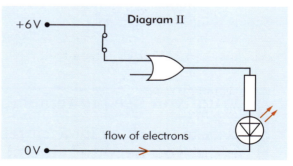

+6V

Diagram II

flow of electrons

0V

What you need to remember [Copy and complete using the **key words**]

How to connect input sensors 1: switches

When a switch is used as an input sensor, it is placed between the _____ side of a low
voltage supply and an _____ to a logic gate.
When the switch is on, the input to the logic gate is then on (or 1, or _____).

10

How to split voltages: the potential divider

Connecting switches into electronic control circuits as input sensors is easy. All you have to do is place them between the positive side of the low voltage supply and an input to a logic gate.

But connecting input sensors such as thermistors and LDRs into electronic control circuits is more tricky. To understand how to do it, you must first understand the circuit shown in the diagram.

1 What is this circuit called?

■ What happens in a potential divider?

The **resistors** R_1 and R_2 in the potential divider circuit are connected to the low voltage supply in series with each other. This means that the current through them must be exactly the same.

When a current flows through a resistor, the voltage (or potential difference) is given by:

$$\begin{array}{ccc} \text{voltage} & = & \text{current} & \times & \text{resistance} \\ \text{(V, volts)} & & \text{(A, amperes)} & & \text{(Ω, ohms)} \end{array}$$

2 Suppose that R_1 and R_2 have the same resistance. What can you then say about:

(a) the voltage across each resistance?

(b) the size of V_{out}?

(c) Why is the circuit called a potential divider?

The Box shows you how to calculate V_{out} when the resistors R_1 and R_2 have <u>different</u> values.

3 R_1 is replaced by a 250 ohm resistor. Calculate the new value of V_{out}.

Ideas you need from *Electricity*

In a series circuit:

■ the same current flows through each part of the circuit

■ the applied voltage is **shared** between the components that are connected in series.

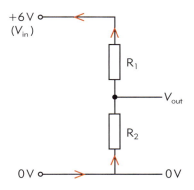

A **potential divider** circuit.

How to calculate V_{out}

$$V_{out} = V_{in} \times \frac{R_2}{R_1 + R_2}$$

Suppose: V_{in} = 6 volts
      R_1 = 2500 ohms
      R_2 = 500 ohms

$$V_{out} = 6 \times \frac{500}{500 + 2500}$$

$$= 6 \times \frac{500}{3000}$$

$$= 1 \text{ volt}$$

What you need to remember [Copy and complete using the **key words**]

How to split voltages: the potential divider

The diagram shows a _____ _____ circuit.
Components R_1 and R_2 are _____.
The voltage supplied (V_{in}) is _____ between R_1 and R_2.

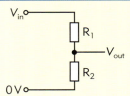

[You should be able to use the formula $V_{out} = V_{in} \times \dfrac{R_2}{R_1 + R_2}$ for the potential divider.]

11 How to connect input sensors 2: LDRs and thermistors

Input sensors such as LDRs and thermistors work because changes in their surroundings make their resistance change.

1 (a) Describe how the resistance of an LDR changes when its surroundings change from dark to light.

   (b) Describe how the resistance of a thermistor changes when you put it into a fridge.

Connecting an LDR into an input sensor circuit

To use an LDR as an input sensor it must be connected as one of the 'arms' of a **potential divider**.

The diagram shows an LDR that is connected in this way.

2 (a) What will happen to the resistance of the LDR as its surroundings become darker?

   (b) What will then happen to the LDR's share of the applied voltage?

   (c) What will then have happened to V_{out}?

Putting the sensor circuit into an electronic control system

When the sensor circuit is used in an electronic control system, V_{out} is used as an input to a **processor** (logic gate).

The graph shows how a processor treats this input.

3 Copy and complete the sentences.

The processor shown on the graph:

■ treats an input below 2.5 volts as being _____

■ treats an input above 2.5 volts as being _____.

So, by choosing an appropriate value for the resistor R, you can make sure that the input to the processor:

■ is 'off' when the sensor is in the light

■ changes to 'on' when the sensor is in the dark.

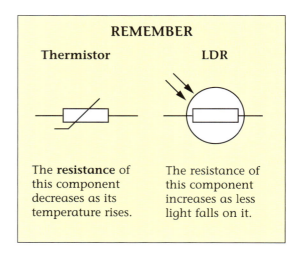

REMEMBER

Thermistor   LDR

The **resistance** of this component decreases as its temperature rises.

The resistance of this component increases as less light falls on it.

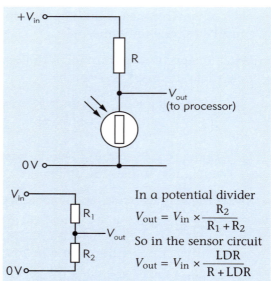

$+V_{in}$

R

V_{out} (to processor)

0 V

V_{in}

R_1

V_{out}

R_2

0 V

In a potential divider

$$V_{out} = V_{in} \times \frac{R_2}{R_1 + R_2}$$

So in the sensor circuit

$$V_{out} = V_{in} \times \frac{LDR}{R + LDR}$$

V_{out} is the LDR's share of the applied voltage (V_{in}).
So as the resistance of the LDR increases compared with R, its share of the applied voltage also increases.

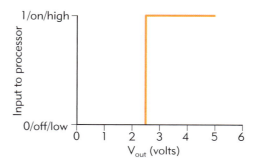

Choosing an appropriate resistor

The diagram shows a temperature sensor circuit.

The graph shows how the resistance of the thermistor in the circuit changes with temperature.

4 What is the resistance of the thermistor at 20°C?

The output from the sensor circuit provides the input to a logic gate.

This input is 'on' for voltages of 3 volts or more.

5 Suppose you want the input to the processor to switch to high when the temperature falls below 20°C.
What value must you choose for the resistance of R?

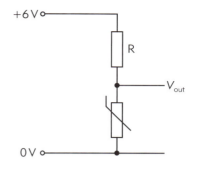

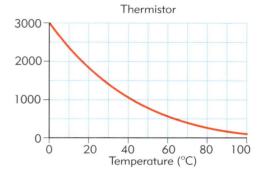

Thermistor

A colour code for resistors

Fixed resistors are quite small components.

They can have many different values.

The Box shows you how to work out the resistance of a fixed resistor from its coloured stripes.

6 Work out the resistances of resistors A, B and C.

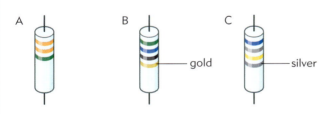

7 How would you identify a 220 Ω (ohm) resistor?

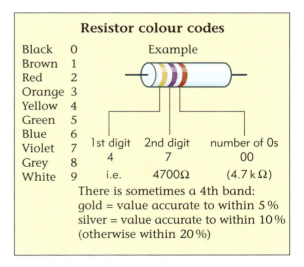

Resistor colour codes

Black   0
Brown   1
Red   2
Orange 3
Yellow 4
Green   5
Blue   6
Violet 7
Grey   8
White   9

Example

	1st digit	2nd digit	number of 0s
	4	7	00
i.e.		4700Ω	(4.7 kΩ)

There is sometimes a 4th band:
gold = value accurate to within 5%
silver = value accurate to within 10%
(otherwise within 20%)

What you need to remember [Copy and complete using the **key words**]

How to connect input sensors 2: LDRs and thermistors

To make input sensor circuits, LDRs and thermistors are used as part of a _____

_____.

The output from the sensor circuit (V_{out}) is used as an input to a _____.
Changes in the surroundings change the _____ of LDRs and thermistors.
This changes the size of _____.
The input to the processor may then change from _____ to _____ (or vice versa).

25

Making a moisture sensor

Water is not a good conductor of electricity.
But, unless it is very pure, it is not a very good
electrical insulator either.
This is why it is dangerous to handle mains electricity
near water or with wet hands.

The diagram shows a simple moisture sensor. It is
made from a small piece of a type of board that is
used for electronic circuits.

1 (a) The dry moisture sensor is connected to the
battery and microammeter.
What reading will there be on the meter?

   (b) Give a reason for your answer to (a).

2 (a) The moisture sensor is then placed in a beaker
of water whilst still connected in the circuit.
What reading will there now be on the meter?

   (b) Use the idea of resistance to explain
your answer.

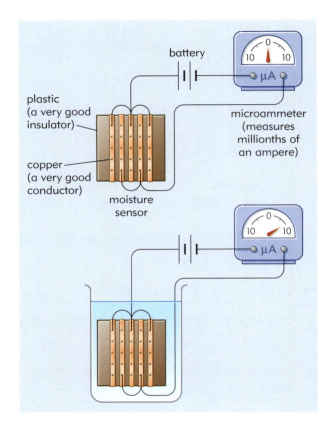

■ A moisture detecting system

The diagram shows how the moisture sensor can be
used in an electronic circuit for controlling the
windscreen wipers on a car automatically.

3 Write down the following statements in the correct
order to explain how the system works.
The first sentence has been chosen for you.

The moisture sensor is dry at first and it has a
resistance much greater than 10 MΩ.

■ V_{out} rises to more than 6 volts.

■ V_{out} is very small.

■ The output from the AND gate goes 'on' and the
windscreen wiper motor starts.

■ The resistance of the moisture sensor falls to less
than 10 MΩ.

■ When it rains the moisture sensor becomes wet.

■ The inputs to the AND gate change to 'on'.

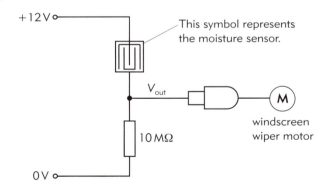

> ### What you need to remember
>
> #### Making a moisture sensor
>
> You should be able to explain, step
> by step, how an electronic control
> system works, just as you have on
> this page.

13 How to adjust input sensor circuits

Diagram I shows an electronically controlled heating system.

The input sensor consists of a thermistor and a resistor in a potential divider circuit.

However, using a <u>fixed</u> resistor in the input sensor circuit isn't very satisfactory.

1 Write down <u>three</u> problems with using a fixed resistor in the input sensor circuit.

■ A more satisfactory input sensor circuit

Diagram II shows how you can make the temperature control of the heating system much easier to set and much more flexible.

2 (a) What is the difference between this control system and the previous one?

  (b) Explain, as fully as you can:

  (i) why it is now much easier to set the input sensor circuit so that the system switches on and off at the required temperature

  (ii) why the system is now more flexible.

3 You want to adapt the system so that it will switch on the cooler in a fridge when the temperature rises above 4 °C.
Suggest <u>two</u> different ways of doing this.

When fixed or variable resistors are used in potential divider circuits, only very small currents flow through them.

So there are no problems caused by them transferring a lot of energy as heat (thermal energy).

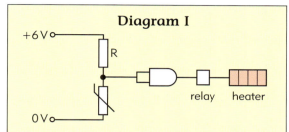

Diagram I

If you want the heating system to switch on and off at a particular temperature, R must have exactly the right value. To find out what this value is you need to know:

■ the resistance of the thermistor at that temperature

■ the exact voltage required for the input to the logic gate to be high.

Furthermore, you can only get fixed resistors with certain resistances (called preferred values). Even then a resistor may be 5 %, 10 % or 20 % different from its colour-coded value.

Finally, with a fixed resistor the electronic thermostat is permanently set at one temperature.

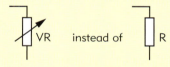

Diagram II

You can **adjust** the **variable** resistor (VR) so that the input to the processor changes from 'off' to 'on' at any **temperature** you choose.

What you need to remember [Copy and complete using the **key words**]

How to adjust input sensor circuits

To make a more convenient and more flexible input sensor circuit, you can use a thermistor (or an LDR) and a _____ resistor in a potential divider circuit.
You can then _____ the circuit so that V_{out} changes from low to high (or vice versa) at whatever _____ (or light level) you choose.

H1 For Higher Tier students only

Putting it all together

Ideas you need from *Electricity*

A diode allows a current to flow through it in one direction only.

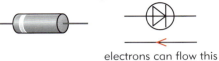

electrons can flow this way through the diode

Here is the full circuit diagram for a light dependent switch.

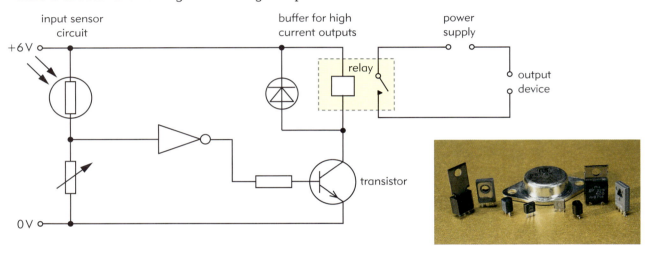

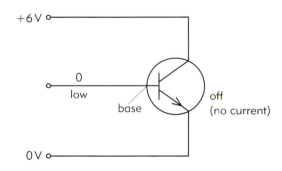

Some transistors.

1 The circuit includes two components which have not previously been referred to in this chapter. What are these components?

The job of each of these additional components will now be considered.

■ What the transistor does

The transistor in this circuit acts as an electronic switch.

The diagrams show when this switch is 'off' and when it is 'on'.

2 Copy and complete the sentence.

When the input to the base of the transistor is high (on), the transistor allows a _____ to flow between its two other terminals.

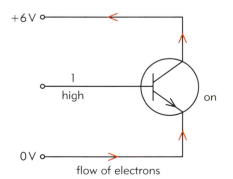

What the diode does

When the current through the coil of the relay is switched off, a potential difference is induced between the ends of the coil.

This potential difference is in the opposite direction to the potential difference supplied to the circuit by the battery. It can damage the transistor.

The diode is included in the circuit to protect the transistor from being damaged.

The diagrams show how the diode does this job.

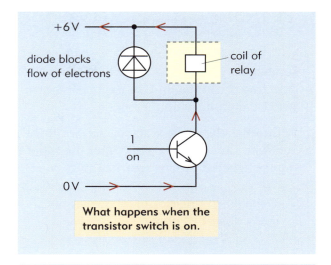

diode blocks flow of electrons

coil of relay

on

What happens when the transistor switch is on.

3 The diode does <u>not</u> affect the circuit when the transistor switch is 'on'.
Explain why, as fully as you can.

4 Copy and complete the sentences.

A diode that is connected so that it does not normally allow a current to flow through it is said to be _____-_____.

When a reverse potential difference is induced across the coil, a current does then flow through the _____.

This prevents this potential difference damaging the _____.

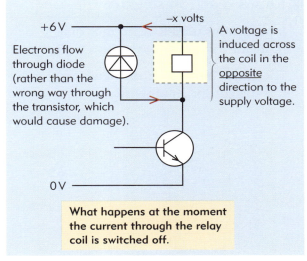

+6V

−x volts

Electrons flow through diode (rather than the wrong way through the transistor, which would cause damage).

A voltage is induced across the coil in the <u>opposite</u> direction to the supply voltage.

0V

What happens at the moment the current through the relay coil is switched off.

The diode is connected so that it does not normally allow a current to flow through it. We say that it is <u>reverse-biased</u>.

What's the delay?

Sometimes you want a delay between a sensor detecting a change and an electronic system switching on an output device.

With other electronic systems, you want the output device to stay on for a while after the input signal that switched it on has returned to zero.

1 Write down <u>one</u> example of each type of time delay described above.

■ Rapid reaction sensors

Input sensors usually react <u>immediately</u> to the changes in their surroundings that they are designed to detect.

The graph shows the rapid reaction of a switch when it is used as an input sensor.

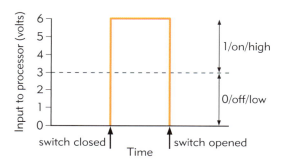

2 Copy and complete the sentences.

The input to the processor goes to on (1, high) immediately the switch is _____.

The input returns to off (0, low) _____ the switch is opened.

■ How to slow down a voltage rise

The diagram shows how you can slow down the rise in the voltage input to a processor when a switch is closed.

3 What <u>two</u> components are used to cause a delay in the voltage rise?

Slowing down the rise in voltage can be used to cause a time **delay** in an electronic control circuit.

To remove the dampness from a shower room, an extractor fan must stay on for a few minutes after the shower has been turned off.

You don't want to set off a burglar alarm every time you go into your house.
So there must be a delay of about 30 seconds between opening the front door and the alarm going off.
This gives you time to key the code into the control box and switch off the alarm system.

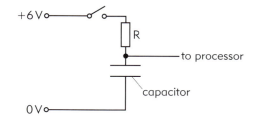

Some capacitors.

Why a capacitor + resistor causes a delay

When the switch in the capacitor + resistor circuit is closed, electrical **charge** flows into the capacitor. As charge flows into the capacitor, the voltage across the capacitor **increases**. This makes it harder for more charge to flow in, so this happens more slowly. Graph I shows how the voltage across the capacitor changes with time as the capacitor **charges**.

4 (a) Describe, in as much detail as you can, how the voltage across the capacitor changes with time.

(b) How long does it take for the voltage across the capacitor to reach half the voltage of the supply?

Graph II is for a different capacitor + resistor circuit connected to the same battery.

5 (a) How quickly does this capacitor charge up compared to the capacitor in Graph I?

(b) Suggest <u>two</u> possible reasons for the difference.

How to delay a voltage fall

If you connect a conductor across a charged capacitor, the charge flows out of the capacitor.
We say that the capacitor **discharges**.

Graph III shows how the **voltage** across a capacitor changes as it discharges through a resistor.

6 (a) Describe, in as much detail as you can, how the voltage across the capacitor changes with time.

(b) How could you make the same capacitor discharge more slowly?

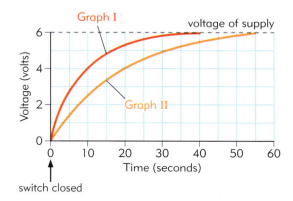

To get a slower charging rate and a slower voltage rise you can:

■ use a **capacitor** with a greater value
■ use a higher **resistance** in series with capacitor.

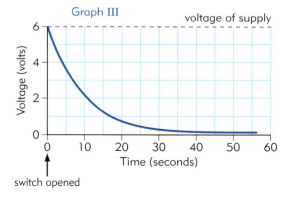

What you need to remember [Copy and complete using the **key words**]

What's the delay?

When a capacitor is connected to a battery, electrical _____ flows into the capacitor.
As the capacitor _____, the voltage across the capacitor _____.
When a conductor is connected across a charged capacitor, electrical charge flows out of the capacitor and _____ it. As this happens, the _____ across the capacitor falls.
To make the voltage rise (or fall) more slowly as a capacitor charges (or discharges), you can use a _____ with a greater value or you can connect a higher _____ in series. Capacitors can be used to produce a time _____ in electronic control circuits.

H2 For Higher Tier students only
More about time delay circuits

Burglar alarm time delay

The diagram shows how a capacitor is connected into the sensor circuit of an electronic burglar alarm system.

Switch 1 is used to switch the whole system on or off (i.e. to enable or disable the system).

To disable the system, the correct code, for example 3274, must be keyed into the control box.

Switch 2 <u>closes</u> whenever the front door of the house is opened.

1 An intruder breaks into the house when the alarm system is enabled (i.e. when Switch 1 is closed). Write down the following sentences in the correct order to explain how the system then operates.

■ The output from the processor goes to high.

■ Switch 2 closes.

■ When the voltage across the capacitor rises to 2.5 volts, input B to the logic gate becomes on (high).

■ The buzzer sounds.

■ An intruder opens the front door.

■ There is a delay whilst the capacitor charges.

2 Write down the following sentences in the correct order to explain how the system operates when the owner of the house returns.

■ Switch 1 opens.

■ Switch 2 closes.

■ Before the voltage across the capacitor rises to 2.5 volts, the owner keys the code into the control box.

■ The owner opens the front door.

■ Nothing else happens.

■ There is a delay while the capacitor charges.

3 (a) Use the graph to find out how long the owner has to key in the code after she has opened the front door.

(b) How can the circuit be adjusted to give the owner a few more seconds to key in the code?

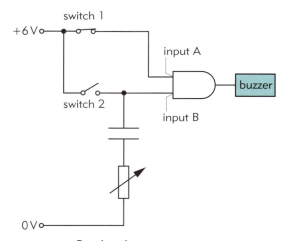

Burglar alarm system

[Actually, there may be lots of other switches, or other sensors, also connected to input B of the AND gate. But to understand how the system works we only need to consider <u>one</u> of these inputs.]

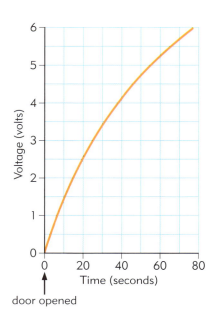

Shower extractor fan time delay

The diagram shows the electronic circuit controlling the extractor fan in a shower room.

When you lock the shower room door, you also close the switch in the electronic control circuit.

When the switch closes, the capacitor becomes charged.

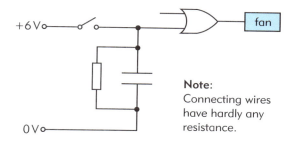

Note: Connecting wires have hardly any resistance.

4 (a) How long does it take for the capacitor to become fully charged?

   (b) Explain your answer to (a).

   (c) Describe, as fully as you can, what you would see (or hear) happening when you lock the shower door.

When you unlock the shower door, you also open the switch in the electronic control circuit.

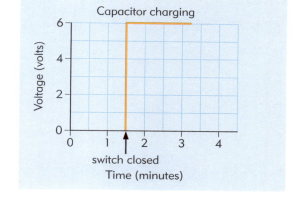

5 (a) Describe, as fully as you can, what then happens to the voltage across the capacitor.

   (b) Explain your answer to (a).

The input to the OR gate is 'off' if it falls to below 2 volts.

6 (a) What happens to the extractor fan when the input to the OR gates goes to 'off'?

   (b) How long does it take, after the switch is opened, for this to happen?

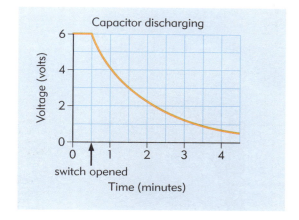

Using your knowledge

1 A door is fitted with a magnetic switch for a burglar alarm circuit. This type of switch normally closes when the door is closed and opens when the door is opened.

Draw a diagram to show how you would modify the burglar alarm system for this type of switch.

2 The extractor fan in a shower room stays on for far too long after the shower room door is unlocked.

Suggest two ways of modifying the circuit to solve this problem.

H3 For Higher Tier students only

Are electronic systems a good thing?

Electronic systems are very versatile and very reliable.
Complex circuits containing many components can be
produced very cheaply in the form of silicon chips.
So electronic circuits are used for more and more different things.

Some examples of things that use electronic circuits are
illustrated below.

Shopping malls, banks, etc. are often monitored
using close circuit television (CCTV).

Many people get cash from a machine using a
plastic card and their personal identification
number (PIN).

During the 1990s there was an enormous
increase in the number of people who have
mobile phones.

In the past 10 years there has been a huge
increase in the use of the Internet (including
e-mails).

As with any other technology, the things that we do
with electronics have advantages and disadvantages.

1 From the list on the opposite page, find at least two
advantages and two disadvantages of each of the
examples shown in the illustrations.
[Some of the items apply to more than one example.]

A They mean that banks can reduce their wage bills considerably.

B There's plenty of interesting work designing web-sites.

C They make it more likely that people committing robberies and assaults will be prosecuted.

D They mean that there are far fewer jobs for workers who meet customers face to face.

E It is difficult to prevent young children accessing unsuitable material.

F They mean that we never know when we are being watched, or by whom.

G They can be a nuisance to other people on buses, on trains, in the cinema and even in school classes.

H They mean that you can access your money at any time of day or night.

I They mean that you can contact your friends wherever you — or they — are.

J They mean that our privacy is invaded.

K They mean that there is a small number of very interesting and well-paid jobs designing the hardware (e.g. the chips containing the electronic circuits).

L They divide people even more into the 'haves' (who can afford them and know how to use them) and the 'have-nots' (who cannot afford them or who don't know how to use them).

M They mean that you can find out information about almost anything without leaving your desk.

N They mean that it can sometimes be difficult for people to contact members of your family on the phone.

O They create jobs for operatives and technicians on the construction lines making the hardware.

P They mean that you can obtain help when you need it wherever you are, for example if your car has broken down.

Q You can send information anywhere in the world, usually within a few seconds.

S They provide security for people in public places.

R They may damage your health, especially if you use them a lot.

Using your knowledge

1 (a) Give <u>two</u> further examples of the use of electronics.

(b) Give <u>two</u> advantages and <u>two</u> disadvantages for each of your examples.

2 Think of <u>one</u> further advantage and one further disadvantage for each of the four examples illustrated on page 34.

Two types of lens

We use lenses in many ways.
The diagram shows some of the optical devices that use lenses.

1 What sort of materials are lenses made from?

2 What can you say about the shape of most lenses?

Lenses are divided into two types depending on what they do to the light that passes through them.

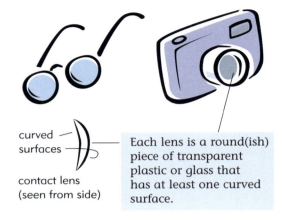

curved surfaces

contact lens (seen from side)

Each lens is a round(ish) piece of transparent plastic or glass that has at least one curved surface.

■ What do lenses do to light?

Lenses change the direction of the light that passes through them.

The diagrams explain how you can use parallel rays of light to sort lenses into two different types.

3 Copy and complete the table.

Type of lens	What it does to parallel rays of light
converging	makes them _____
_____	_____

■ Looking at lenses

When you look at them from the side, lenses can have many different shapes (see diagrams below). But all lenses are either converging or diverging.

The diagrams show what happens when parallel rays of light pass through six different shaped lenses A–F.

4 (a) Which of lenses A–F are converging lenses and which are diverging lenses?

(b) What is the same about the shape of <u>all</u> of the converging lenses that makes them different from the diverging lenses?

You can use a ray box to produce narrow beams of light. These narrow beams of light are called rays.

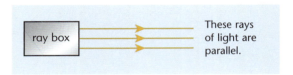

ray box

These rays of light are parallel.

This lens makes parallel rays of light come to a point (converge).
So we call it a **converging** lens.

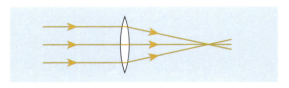

This lens makes parallel rays of light spread out (diverge).
So we call it a **diverging** lens.

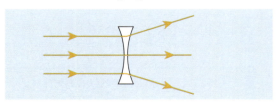

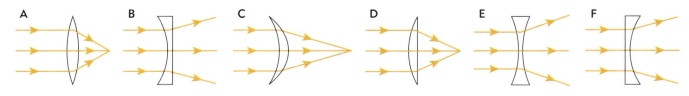

A   B   C   D   E   F

What is the focus of a lens?

A converging lens makes parallel rays of light come together at a point.

This point is called the **focus** of the lens.

5   A diverging lens also has a focus (see diagram).
    Write two sentences, similar to the ones above, about
    a diverging lens.

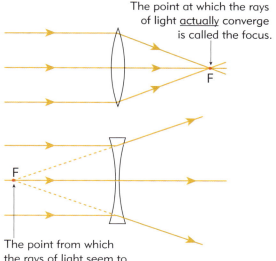

The point at which the rays of light <u>actually</u> converge is called the focus.

The point from which the rays of light <u>seem</u> to diverge is called the focus.

What happens to rays of light that aren't parallel?

Rays of light may be spreading out (diverging) before they pass through a converging lens.

The lens may not then be strong enough to make the rays of light converge to a point (see diagram).

6   What effect does a converging lens have on
    diverging rays of light if it can't make them
    converge to a point?

7   Where would you need to put the lamp so that a
    parallel beam of light comes out from the lens?
    [Hint: think what would happen if a parallel beam of
    light was travelling in the opposite direction.]

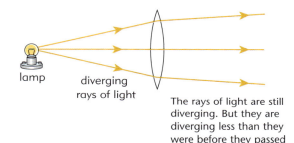

lamp   diverging rays of light

The rays of light are still diverging. But they are diverging less than they were before they passed through the lens.

Rays of light may be coming together (converging) before they pass through a diverging lens.

The lens may not then be strong enough to make the rays of light diverge (see diagram).

8   What effect does a diverging lens have on converging
    rays of light if it can't make them diverge?

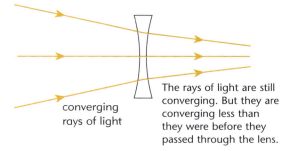

converging rays of light

The rays of light are still converging. But they are converging less than they were before they passed through the lens.

What you need to remember [Copy and complete using the **key words**]

Two types of lens

When parallel rays of light pass through a _____ lens they come together at a point.
When parallel rays of light pass through a _____ lens they spread out as if they had all come from a point.
In both cases we call this point the _____ of the lens.

[You should be able to draw what happens to parallel rays of light when they pass through both types of lens.]

Two types of image

The drawing shows how you can use a converging lens to produce a picture of a lamp on a screen.

lamp
(object)

converging
lens

screen
(e.g. sheet of paper)

image

The picture on the screen is called an **image**.

In this case, it is light from the lamp that produces the image. So we say that the lamp is the **object**.

1 (a) How does the size of the image compare with the size of the object?

(b) Write down <u>one</u> other difference that you can see between the image and the object.

> **Another word for upside-down**
>
> When an image is the opposite way up from the object we say that the image is <u>inverted</u>.

■ Forming an image of a distant object

You can use a converging lens to produce an image of a distant object on a screen.

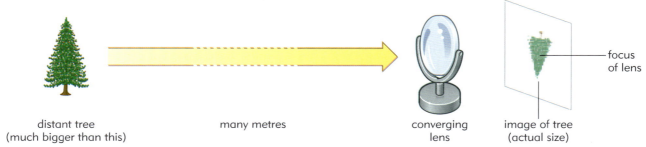

distant tree
(much bigger than this)

many metres

converging
lens

image of tree
(actual size)

focus
of lens

2 (a) Where must you place the screen to get a sharp image of a distant object using a converging lens?

(b) How does the distance of the image from the lens compare with the distance of the object?

3 The image of the distant object produced by the converging lens is called a <u>real</u> image.
Explain why.

To get a sharp image, the centre of the screen must be at the focus of the lens. This means that the image is much **nearer** the lens than the object is.
An image that you can see on a screen is called a **real** image.

How a camera works

A simple camera consists of:

- a converging lens at the front of a light-proof box

- a **film** instead of a screen

- a shutter which lets in light through the lens for a fraction of a second when you press the button.

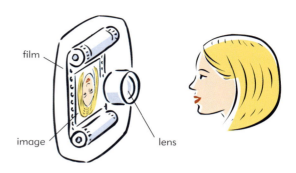

film

image

lens

4 (a) Where must the film be placed so that you get a sharp image of distant objects?

(b) What must you be able to do if you want a clear image of an object that is near to the camera?

In a cheap camera, the position of the lens is fixed so that you get a sharp image on the film of distant objects (i.e. things more than about 3 metres away).

With more expensive cameras you can get a sharp image of things closer than 3 metres by moving the lens slightly further away from the film.

Another type of image

When you look at an object through a diverging lens you can see an image.

object

The object looks <u>as if</u> it is here. But you <u>can't</u> get the image on a screen. So we call it a **virtual** image.

eye

5 Write down:

(a) <u>one</u> way in which the image of a distant object produced by a diverging lens is the <u>same</u> as the image in a camera;

(b) <u>three</u> ways in which the images are <u>different</u>.

What you need to remember [Copy and complete using the **key words**]

Two types of image

You can use a converging lens to form an _____ of an object on a screen.
An image that you can form on a screen is called a _____ image.
An image that you cannot form on a screen is called a _____ image.
In a camera, a converging lens is used to form an image on a _____.
The image formed by a camera is:

- smaller than the _____

- _____ to the lens than the object.

H4 For Higher Tier students only

More about real images

The diagram shows how you can use a scale drawing to find the exact size and position of a real image produced by a converging lens.

You begin by drawing the lens and its main axis.
Then you mark, to scale, on the axis:

- the position of the object (in this case 80 cm)

- the position of the focus on each side of the lens (in this case 25 cm; see 'Focal length' Box).

REMEMBER

When parallel rays of light pass through a converging lens they converge at the focus of the lens.

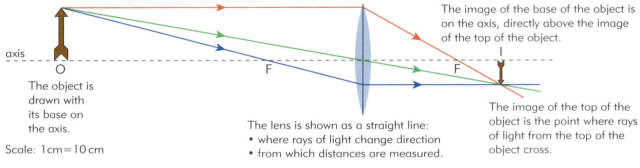

The image of the base of the object is on the axis, directly above the image of the top of the object.

axis

The object is drawn with its base on the axis.

Scale: 1 cm = 10 cm

The lens is shown as a straight line:
• where rays of light change direction
• from which distances are measured.

The image of the top of the object is the point where rays of light from the top of the object cross.

When a real image is formed, every ray of light from a particular part of the object that passes through the lens must end up at the same point on the image.

You know what happens to <u>three</u> of the rays from the top of the object when they pass through the converging lens:

- the red (top) ray is parallel to the axis and passes through the focus on the opposite side of the lens

- the green (middle) ray passes through the centre of the lens and goes straight on

- the blue (bottom) ray passes through the left focus of the lens and so is parallel to the axis <u>after</u> it has passed through the lens.

You find where the image of the top of the object is by drawing <u>any</u> <u>two</u> of these rays.
[You can then draw the third ray to check that you haven't made a mistake with the first two.]

Focal length

Every lens has what is called a focal length.
This is the distance between the centre of the lens and its focus.

Magnification

The magnification produced by a lens is

$$\frac{\text{size of image}}{\text{size of object}}$$

When the image is smaller than the object, the magnification is a fraction, e.g. $\frac{1}{2}$ or 0.5.

1 (a) How far away from the lens is the image

    (i) on the diagram? (ii) in fact, taking into account the scale?

  (b) How big is the image compared to the object?

  (c) What is the magnification when the object is in this position?

2 (a) Draw a similar diagram for an object that is 120 cm from the same lens.

  (b) Repeat question 1(a)–(c) for this new position.

The image of a distant object

The converging lens in a camera (or in your eye) often forms an image of a distant object.

The rays of light that reach a lens from each point on a distant object are almost parallel.

Diagram I tells you where the image will be, but it does not show how you get a real image.

3 (a) Where does a converging lens form an image of a distant object?

(b) How far away from the lens will this image be?

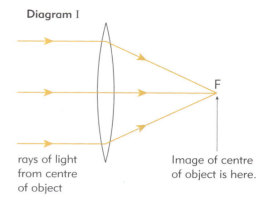

Diagram I

rays of light
from centre
of object

Image of centre
of object is here.

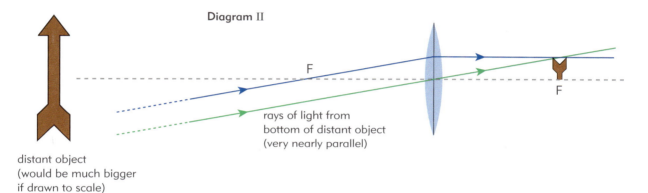

Diagram II

F

rays of light from
bottom of distant object
(very nearly parallel)

F

distant object
(would be much bigger
if drawn to scale)

Diagram II shows how a converging lens uses the light from the bottom of a distant object to form an image.

4 Make a careful copy of the diagram.
Then complete it by showing what happens to two parallel light rays from the top of the object.

There are two rays that you can draw:

- the green (lower) ray passes through the centre of the lens and goes straight on

- the blue (upper) ray passes through the left focus of the lens and is parallel to the axis after it has gone through the lens.

Using your knowledge

1 (a) Draw a diagram to find the size and position of the image produced by a converging lens of focal length 20 cm when the object is placed 25 cm from the lens.
[Hints: Use a scale of 1:10. Place the object close to the left of your paper.]

(b) What magnification does the lens produce when the object is in this position?

(c) Calculate the value of

$$\frac{\text{distance of image}}{\text{distance of object}}$$

What do you notice about this value?

2 By placing the object <u>just a little</u> further away from a converging lens than its focus, you can produce a <u>greatly magnified</u>, real image.
What optical device makes use of this idea?

H5 For Higher Tier students only

Bigger – but not real!

◼ How does a magnifying glass work?

You can use a converging lens as a magnifying glass.

The diagram shows how an image is produced when a converging lens is used in this way.

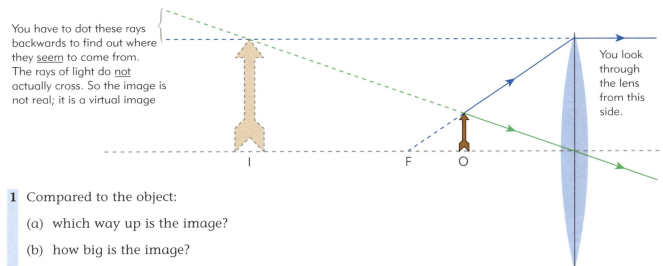

You have to dot these rays backwards to find out where they <u>seem</u> to come from. The rays of light do <u>not</u> actually cross. So the image is not real; it is a virtual image

You look through the lens from this side.

I   F   O

1 Compared to the object:

   (a) which way up is the image?

   (b) how big is the image?

2 (a) Why is a converging lens used in this way called a magnifying glass?

   (b) Where must the object be placed to be magnified in this way?

   (c) Write down <u>two</u> reasons for saying that the image produced by a magnifying glass is a <u>virtual</u> image.

◼ What happens if the object is at the focus?

When an object is placed at the focus of a converging lens, the rays of light from each part of the object are parallel after they have passed through the lens.

In other words:

◼ the rays of light do not <u>actually</u> cross at a point

◼ the rays of light do not <u>appear</u> to come from a point.

3 Explain why a converging lens forms neither a real image nor a virtual image of a small object placed at its focus.

Even though the lens does not form an image of an object placed at the focus, this is the best position for the object when the lens is used as a magnifying glass. To explain why, you need to understand how your eye forms an image.

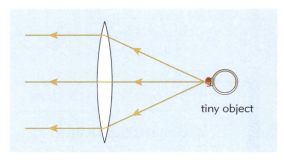

tiny object

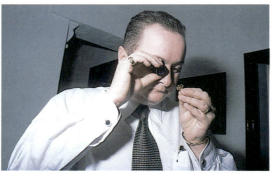

Your eye as an optical instrument

The diagrams show how your eyes form images of distant and near objects on the retina.
Information from each retina is then sent along your optic nerves to your brain.

4 Many people think that it is the lens in your eye that produces an image on the retina.
Explain why this is not completely true.

5 Your eye is more relaxed when you are looking at a distant object than when you are looking at a near object. Explain why.

6 So far as your eye is concerned, looking at a tiny object that is at the focus of a magnifying glass is exactly the same as looking at a distant object. Explain why.

Normally, you can't see an object clearly if you bring it nearer than about 25 cm to your eye.

But with a magnifying glass you can bring an object much closer to your eye and still see it clearly.

This means that the object looks much bigger (see Box).

Distance and apparent size

The nearer an object is to your eye, the bigger the image it produces on your retina. So the bigger the object looks.

For example, if you move an object so that it is only half as far from your eye, it will look twice as big.

---

7 Suggest why you can't normally see an object clearly if it is too close to your eye.

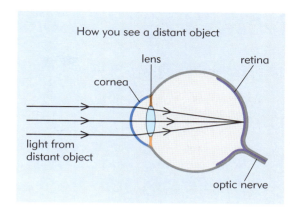
How you see a distant object

The almost parallel rays of light from each part of a distant object change direction when they pass through the cornea and the lens. A real image is formed on the retina.

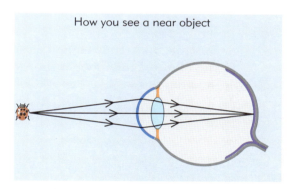
How you see a near object

To produce an image of a near object on your retina, the ciliary muscles in your eye have to squeeze the lens to make it more curved. There is a limit to how much they can do this.

Using your knowledge

1 A botanist is looking at the structure of a small flower. The nearest she can bring the flower to her eye and still see it clearly is 30 cm.

How many times bigger can she make the flower look:

(a) using a hand lens with a focal length of 10 cm?

(b) using a hand lens with a focal length of 5 cm?

[See page 40 for a reminder about focal length.]

Turning forces

The pushchair shown on the diagram is free to move.

When an unbalanced force acts on the pushchair its movement changes.

> **1** Write down the different ways that the movement of the pushchair can change when an unbalanced force acts on it.

The roundabout shown on the diagram is mounted on a post that is fixed to the ground.

So when you apply a force to the roundabout it does not move <u>along</u>.

But a force can still make the roundabout move.

> **2** (a) Explain <u>in terms of forces</u> why the roundabout cannot move along.
>
> (b) Describe, in as much detail as you can, the way that the roundabout shown on the diagram moves.

The roundabout rotates if an unbalanced **turning** force acts on it.

■ **How to increase
the turning effect of a force**

To undo a bolt you need to apply a turning force.

Unless the bolt is very slack, you cannot produce a big enough turning force using your fingers.

To produce a big enough turning force you need to use a spanner.

Sometimes you cannot produce a big enough turning force to undo a bolt even with a spanner.

The diagram shows how you can solve this problem.

> **3** (a) How can you make the same force produce a bigger turning effect?
>
> (b) Why do you think this works?

Ideas you need from *Forces*

When an unbalanced force acts on an object, it can make the object:

■ accelerate (i.e. start moving and/or speed up)

■ slow down (and possibly stop moving)

■ change direction.

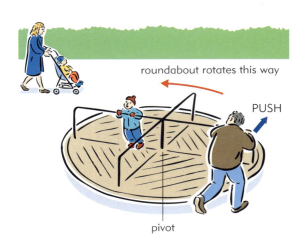

roundabout rotates this way

PUSH

pivot

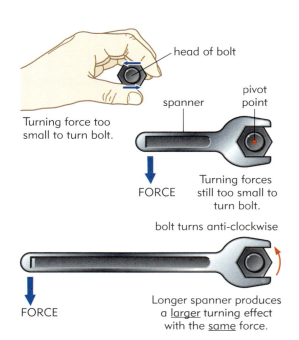

head of bolt

Turning force too small to turn bolt.

spanner

pivot point

FORCE

Turning forces still too small to turn bolt.

bolt turns anti-clockwise

FORCE

Longer spanner produces a <u>larger</u> turning effect with the <u>same</u> force.

Comparing turning forces

To compare turning forces, you need some way of measuring their turning effect. The Box explains how you can do this.

4 (a) What two factors affect the turning effect of a force?

(b) How can these two factors be used to calculate the turning effect of a force?

(c) What other name is used for the turning effect of a force?

(d) What unit is used for the turning effect or moment of a force?

Calculating moments

The diagrams below show some turning forces.

5 (a) Calculate the moment of each force.

(b) For each force state whether its moment is clockwise or anti-clockwise.

(c) Will the see-saw rotate clockwise or anti-clockwise? Give a reason for your answer.

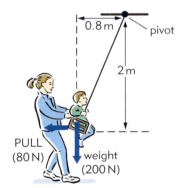

To produce a bigger turning effect you can:
■ use a bigger force
■ increase the distance between the line of action of the force and the pivot point.

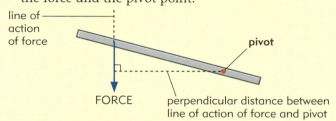

line of action of force

FORCE

pivot

perpendicular distance between line of action of force and pivot

This distance is always measured at right angles to, i.e. **perpendicular** to, the line of action of the force.

So the turning effect or **moment** of a force is calculated like this:

moment	=	force	×	perpendicular distance between
[newton metres, N m]		[newtons, N]		line of action of force and pivot [metres, m]

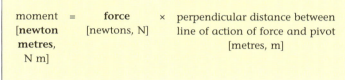

Example   12 cm

25 N force

Moment = 25 N × 0.12 m
       = 3 N m (↶)

pivot

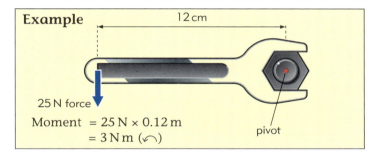

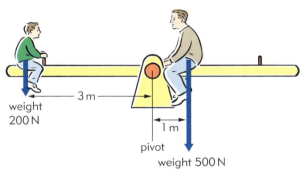

weight 200 N

3 m

1 m

pivot

weight 500 N

What you need to remember [Copy and complete using the **key words**]

Turning forces

A force that is applied to an object at a distance from a pivot produces a _____ effect. This is called the _____ of the force and can be calculated using:

moment   =   _____   ×   _____   distance between line of action of force and _____

[_____ , N m]   [newtons, N]   [metres, m]

45

2

When the swinging stops

The diagram shows a child on a swing.
The swing hangs down from a support.
We say that the swing is <u>suspended</u>.

As it goes backwards and forwards, the swing is in fact rotating about a pivot point. It rotates in one direction for a short time, then in the opposite direction for a short time, and so on.

1 Where is the pivot point of the swing?

To keep the swing going, the child has to keep bending and straightening his legs. Otherwise, the swing eventually stops swinging and comes to rest.

2 What can you say about the position of the child when the swing stops swinging?

Not all of the child's body can be directly below the pivot point when the swing stops.

In fact, when the swing stops it is the child's **centre of mass** that is vertically below the point of suspension.

What is centre of mass?

Every little bit of a child's body has some mass.
Weight is the force of the Earth's gravity acting on a mass.
So every little bit of the child's body has **weight**.

It is more convenient to add up the weights of all the bits of the child's body and think of his <u>total</u> weight as a <u>single</u> force.

But if we want to know about the **turning** effect of his weight we need to know exactly where this single force acts. This point is the child's centre of mass.

3 (a) What is the perpendicular distance between the line of action of the child's weight and the pivot when the swing comes to rest?

(b) Use your answer to (a) to explain why the swing comes to rest with the child's centre of mass vertically below the point of suspension.

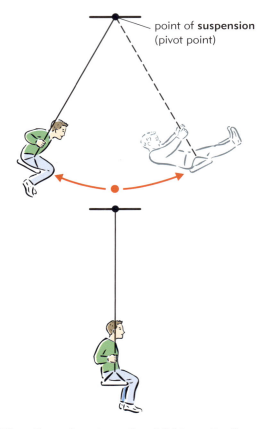

point of **suspension** (pivot point)

When the swing stops, the child is vertically below the point of suspension.

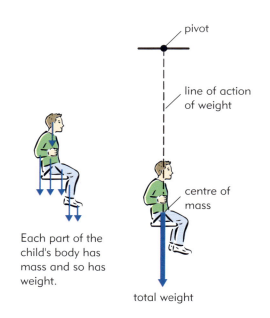

pivot

line of action of weight

centre of mass

Each part of the child's body has mass and so has weight.

total weight

Where is an object's centre of mass?

Some objects can be divided down the middle into two identical halves (see Box). We say that these objects are symmetrical. The centre of mass of a symmetrical object must be on any plane that divides the object into two identical halves.

If a symmetrical object is flat (i.e. cut out from a thin sheet of material) its centre of mass must be on any line that divides it into two identical halves.
Such a line is called an **axis of symmetry**.

If a flat object has at least two axes of symmetry you can easily work out where its centre of mass must be.

4 (a) Choose from the following characters those that have at least two axes of symmetry:

0 1 2 3 4 5 6 7 8 9

A B C D E F G H I J K L M N O P Q R S T U V W X Y Z

   (b) Show, in each case, where the centre of mass is.

Finding the centre of mass of any flat shape

If a flat object has only one axis of symmetry, or none at all, you can still find out where the centre of mass is. The diagram shows one way of doing this.

5 (a) Which of the characters in question 4 have one axis of symmetry?

   (b) Put the sentences opposite in the correct order to describe how to find the centre of mass of a cut-out of one of these characters.

   (c) In what way would finding the centre of mass be different for a character with no axis of symmetry?

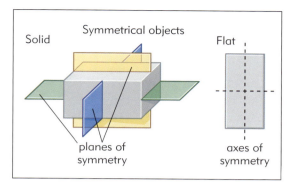

Symmetrical objects
Solid — Flat
planes of symmetry — axes of symmetry

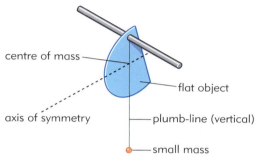

centre of mass — flat object
axis of symmetry — plumb-line (vertical) — small mass

Notes
(i) The centre of mass may be outside the object. If so, you will need to stick the object on to some very thin paper (with hardly any mass).
(ii) If an object has no axes of symmetry, you will need to suspend it from two different holes, mark two vertical lines, then find where these cross.

Use these sentences for question 5(b)

■ Hang a plumb-line from the same point of suspension. Mark a vertical line on the object.

■ Note where the two lines cross.

■ Mark the axis of symmetry of the object.

■ Make a hole very close to the edge of the object and suspend the object from it.

What you need to remember [Copy and complete using the **key words**]

When the swinging stops

The point on a body where you can think of all of its weight acting is called the _____ of _____.

When an object is suspended, it comes to rest with the centre of mass vertically below the point of _____. This means that the _____ of the object does not produce a _____ effect.

The centre of mass of a symmetrical flat object must lie on each _____ of _____.

[You should be able to describe how to find the centre of mass of a thin sheet of any shape.]

H1 For Higher Tier students only

Balanced turning forces

The diagram shows two children on opposite sides of a see-saw.

The see-saw is balanced.
So the weights of the two children must produce turning forces (moments) of the same size but in opposite directions.

1 (a) Make a copy of the diagram.

   (b) Calculate the moment produced by the weight of each child and state its direction (i.e. clockwise or anti-clockwise).

When two turning forces produce no overall turning effect their moments must be balanced.

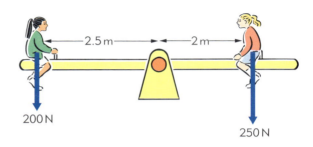

200 N

250 N

2 Copy and complete:

   clockwise moment = anti-clockwise moment
   so the see-saw is _____.

■ Using moments to explain levers

It is difficult to remove the lid from a can of paint using your fingers.
But you can easily remove the lid if you use a screwdriver as a lever.

The diagram shows how you can use the idea of moments to explain how the lever works.

3 (a) Make a copy of the diagram.

   (b) Calculate the anti-clockwise moment of the force applied to the handle of the screwdriver.

   (c) What clockwise moment would the lid of the tin need to provide to balance the anti-clockwise moment?

   (d) What force would the lid need to exert on the tip of the screwdriver to produce this moment?

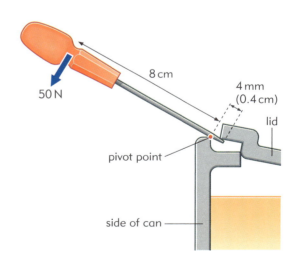

In fact, the friction between the lid and the can cannot provide enough force for the moments to balance.
So the screwdriver rotates, with the edge of the can as the pivot point, and the lid comes off.

What if there are more than two turning forces?

However many turning forces are acting, the same basic idea still applies:

sum of clockwise　=　sum of anti-clockwise
moments　　　　　　　　moments

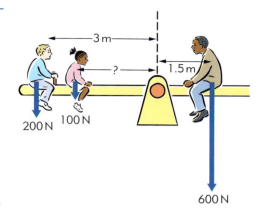

4 (a) Make a copy of the diagram.

(b) Calculate the clockwise moment of the adult.

(c) Calculate the anti-clockwise moment of the left-hand child.

(d) Calculate the anti-clockwise moment that has to be provided by the weight of the second child for the see-saw to be balanced.

(e) Calculate how far from the pivot point the second child has to be to produce this moment.

Taking the weight of a lever into account

A see-saw pivots in the middle, i.e. along one of its axes of symmetry. This means that the centre of mass is at the pivot point so the weight of the see-saw does not have any turning effect.

But in the simple machine shown on the diagram, the lever is not pivoted at its centre of mass.
So we have to take the weight of the lever into account when we calculate the turning forces acting on the lever.

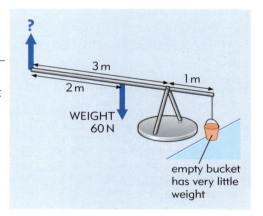

empty bucket has very little weight

5 (a) Make a copy of the diagram for when the bucket is full of water.

(b) What is the anti-clockwise moment produced by the weight of the lever?

(c) What clockwise moment is produced by the bucket when it is full of water?

(d) What anti-clockwise moment is needed, in addition to that produced by the weight of the lever itself, to balance the clockwise moment?

(d) What downwards force must be applied to the lever to produce the additional anti-clockwise moment?

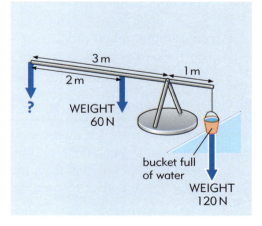

bucket full of water

WEIGHT 120 N

Using your knowledge

1 The biggest load a gardener can lift off the ground is 500 N. But she can lift 1000 N load in a wheelbarrow. Explain why.

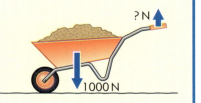

H2 For Higher Tier students only

Why do things fall over?

The diagrams show what happens if you stand a brick on its end and then tilt it.

1 Describe, as fully as you can, what the diagrams show.

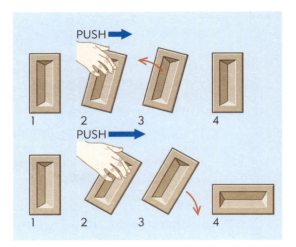

■ Why does (or doesn't) the brick fall over?

When the brick is tilted, it pivots along one edge. This means that the weight of the brick has a turning effect on the brick.

Whether the brick falls over or not depends on where the centre of mass of the brick is in relation to the pivot point.
The diagrams show why this matters.

2 Copy and complete the sentences.

If the brick isn't tilted very far, the line of action of its _____ is to the _____ of the pivot edge.

So the weight produces an _____ moment.

This moves the brick moves back to its _____ position.

If the brick is tilted far enough, the _____ of _____ of its weight is to the right of the _____ edge. So the weight produces a _____ moment. This makes the brick _____ _____.

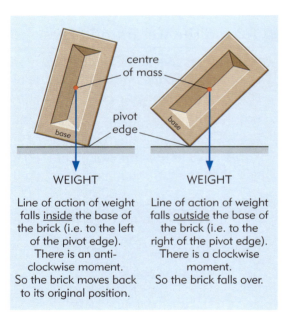

Line of action of weight falls <u>inside</u> the base of the brick (i.e. to the left of the pivot edge). There is an anti-clockwise moment. So the brick moves back to its original position.

Line of action of weight falls <u>outside</u> the base of the brick (i.e. to the right of the pivot edge). There is a clockwise moment. So the brick falls over.

■ Why the brick is now more stable

You can make the brick fall over again if you push it from the side. But you have to tilt the brick more to make the brick fall over from this position.
So we say that the brick is more **stable**.

3 (a) Draw diagrams of a brick on a flat surface:
   (i) standing vertically
   (ii) lying horizontally.

   (b) Give <u>two</u> reasons why the brick that is lying horizontally is more stable than the brick that is standing vertically.

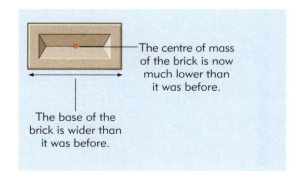

The centre of mass of the brick is now much lower than it was before.

The base of the brick is wider than it was before.

Comparing stability

An object falls over if the line of action of its weight falls outside its base line.

This is less likely to happen, and so the object will be more stable:

- if the centre of mass of the object is low
- if the object has a wide base.

4 Draw glasses A–C and say which is the most stable and which is the least stable.
Explain your answers as fully as you can.

5 (a) Explain why a person is less stable than a pig when they are both standing up.

(b) Suggest how a person who is standing up can make it harder for someone to push them over

(i) from the side   (ii) from the front or back.

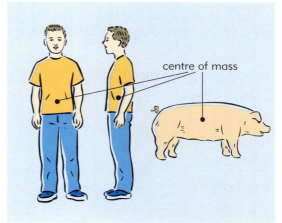

centre of mass

Making sure that buses are stable

Double-decker buses have to pass a stability test to make sure that they cannot topple over too easily. The diagram shows how the stability test is done.

6 (a) Why are double-decker buses likely to be less stable than most other vehicles?

(b) What can designers of the buses do to make them as stable as possible?

(c) How do the testers know whether the bus would fall over if it was tilted to an angle of 15°?

(d) Why are sacks of sand put on the seats in the upper deck of the bus only?

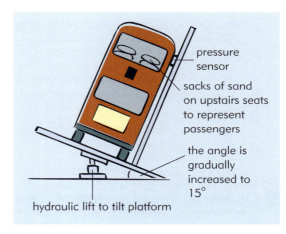

pressure sensor

sacks of sand on upstairs seats to represent passengers

the angle is gradually increased to 15°

hydraulic lift to tilt platform

Using your knowledge

1 The diagrams show a pencil in three different positions on a table. Explain why it is easy to place the pencil in position A, more difficult to place it in position B and impossible to place it in position C.

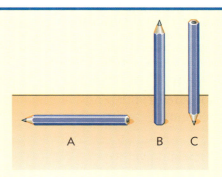

How to accelerate without going faster

The picture shows a car waiting at some traffic lights.

When the traffic lights turn to green, the car accelerates away.

The Box shows how you can calculate the acceleration of the car.

1 (a) The car is accelerating because its speed is changing. But the formula in the Box does <u>not</u> use the idea of speed. What idea does it use instead?

(b) Explain the difference between speed and velocity.

How to calculate acceleration

$$\text{acceleration (metres per second squared, m/s}^2) = \frac{\text{change in velocity (metres/second, m/s)}}{\text{time taken for change (seconds, s)}}$$

If the car accelerating from the traffic lights reaches a speed of 22 m/s in 11 seconds its acceleration is

$$\frac{22}{11} = 2 \, \text{m/s}^2$$

Note: The velocity of a moving object is the speed of the object <u>and its direction</u>.

When the traffic lights turn to green, the car in the picture accelerates away in a straight line. It keeps moving in the same direction. This means that the change in its **velocity** and the change in its speed are exactly the same thing.

Changing velocity without changing speed

The velocity of an object involves both its speed <u>and its direction</u>. This means that the velocity of an object can change even though its speed remains the same.

The diagram shows an example of this.

2 Explain how the <u>velocity</u> of the rubber bung can be constantly changing even though its <u>speed</u> does not change.

When an object moves in a circle, the direction of its movement is constantly changing. This means that the velocity of the object is also constantly changing. So the object must also be **accelerating** all the time.

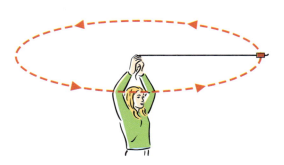

The rubber bung is being whirled round in a horizontal circle on the end of a piece of string. It makes three complete revolutions each second.

What makes an object move in a circle?

An object that moves in a circle is accelerating all the time. To produce an acceleration you need a **force**. So when an object moves in a circle there must be a force acting on it all the time.

3 Look at the diagram of a rubber bung being whirled in a horizontal circle on the end of a piece of string.

   (a) What provides the force that is needed to keep the bung moving in a circle?

   (b) In what direction is this force?

   (c) In what direction does the bung accelerate?

To keep an object moving in a circle at a steady speed there must be a constant force acting on the object towards the centre of the circle. This force is called a **centripetal** force.

4 What happens to the bung if the string breaks? Explain your answer.

What affects the size of the centripetal force?

The size of the centripetal force needed to keep an object moving in a circle depends on:
- the **mass** of the object   ■ the **radius** of the circle
- the **speed** of the object.

The graphs show how the size of the centripetal force is affected by each of these three factors.

5 Describe what each of the graphs shows.

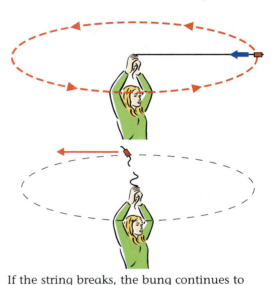

■⟵ = force (tension in string)

If the string breaks, the bung continues to move as shown. We say that it flies off <u>at a tangent</u> (to the circle).

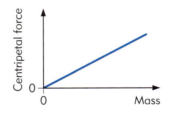 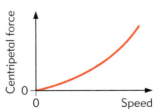 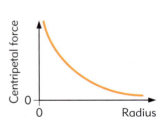

What you need to remember [Copy and complete using the **key words**]

How to accelerate without going faster

When an object moves at a steady speed in a circle, its _____ is constantly changing. This means that the object is _____ all the time.

To produce this constant acceleration, a _____ is needed, acting towards the centre of the circle.

This is called a _____ force and it needs to be greater:
- the greater the _____ of the object   ■ the higher the _____ of the object
- the smaller the _____ of the circle.

Circular motion 1: in the solar system and beyond

Astronomical objects often move around others in approximately **circular** orbits.

The diagrams show some of the astronomical objects that move in this way.

Billions of stars, including the Sun, move in circular paths around the centre of the Milky Way galaxy.

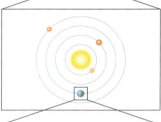

The **planets** move in approximately circular orbits around the Sun.

The **Moon** moves in an approximately circular orbit around the Earth.

1 Write down <u>three</u> examples of astronomical objects that move approximately in a circle.

2 What supplies the centripetal force needed for astronomical bodies to keep moving in this way?

3 It is not quite correct to say that the planets move in circular orbits around the Sun. Explain why. [Higher Tier students should give <u>two</u> reasons.]

REMEMBER

To keep an object moving in a circle, a constant **centripetal** force must act on the object towards the centre of the circle.

Ideas you need from *Forces*

There is a force of attraction between any two objects because of their mass. This force is called **gravity**.

The orbits of planets are not quite circular. They are elliptical. Except for Pluto, however, their orbits are <u>almost</u> circular.

circle ⃝   ⬭ ellipse

For Higher Tier students only

The Moon does <u>not</u>, in fact, orbit the Earth. The Moon and the Earth both orbit the centre of mass of the Earth–Moon system.

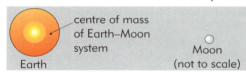

centre of mass of Earth–Moon system

Earth

Moon (not to scale)

Because the Earth is much more massive than the Moon, the centre of mass of the Earth–Moon system is inside the Earth. So it is <u>almost</u> true to say that the Moon orbits the Earth.

A similar argument applies to the Earth and the Sun.

What you need to remember [Copy and complete using the **key words**]

Circular motion 1: in the solar system and beyond

Astronomical objects often move in approximately _____ orbits.

For example, the _____ move around the Sun and the _____ moves around the Earth.

To keep an object moving in a circle, a _____ force is needed.

For astronomical objects this force is supplied by _____.

[You should be able to interpret information about circular motion in the way that you have on this page.]

5

Circular motion 2: inside atoms

An atom consists of a small, massive nucleus containing protons and neutrons.

Electrons move about in the space around the nucleus.

What atoms are made from		
Particle	Mass (relative)	Electrical charge (relative)
proton	1	+1
neutron	1	0
electron	1/1840	−1

The diagram shows a helium atom.

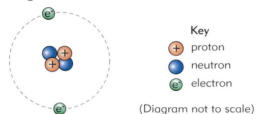

Key
+ proton
● neutron
e⁻ electron

(Diagram not to scale)

1 Describe the helium atom as fully as you can.

Electrons move around the nucleus of an atom in approximately **circular** paths.
These are called orbitals.

2 (a) What supplies the centripetal force needed to keep electrons moving in circular paths?
Answer in as much detail as you can.

(b) What do you think would happen to the electrons near to the nucleus of an atom if they stopped moving?
Give a reason for your answer.

REMEMBER

To keep an object moving in a circle, a constant **centripetal** force must act on the object towards the centre of the circle.

Ideas you need from *Electricity*

There are two types of electrical charge:

positive (+)   and   **negative** (−)

Like charges (i.e. two positive or two negative) repel.

Unlike charges (i.e. a positive and a negative) attract.

Key
  ➤ = Force

The forces between electrically charged objects are called **electrostatic** forces.

For Higher Tier students only

The mass of an electron is only about 1/2000 of the mass of a proton or a neutron.

So we can think of electrons orbiting the nucleus of an atom just like we can think of the Moon orbiting the Earth or the Earth orbiting the Sun (see page 54).

What you need to remember [Copy and complete using the **key words**]

Circular motion 2 : inside atoms

Electrons move in approximately _____ orbitals around the nucleus of an atom.
To keep an object moving in a circle, a _____ force is needed.
For the electrons in an atom, this force is supplied by the _____ force of attraction between _____ and _____ charges.

[You should be able to interpret information about circular motion in the way that you have on this page.]

Circular motion 3: spin driers

The diagrams show what happens when a washing machine spins clothes to make them drier.

1 Copy and complete the sentences.

The drum _____ very quickly.

A _____ force is needed to keep the wet clothes inside the drum moving in a circle.

This force is provided by the inside of the _____.

Where there are holes in the drum, the drum cannot supply any centripetal force.

The holes are too _____ for the clothes to go through.

The _____ from the clothes flies out of the holes at a _____ to the drum. It is then pumped away.

The faster the drum spins, the drier the clothes will be after they have been spun.

2 Washing machine A spins at 1100 rpm.
Washing machine B spins at 1300 rpm.

(a) Explain what these figures mean.

(b) Which washing machine will be better at spinning clothes dry? Give a reason for your answer.
[You can assume that the washing machines have drums with the same radius.]

For Higher Tier students only

3 Two washing machines, C and D, both spin at 1200 rpm. The drum of washing machine C has 1.25 times the radius of the drum of washing machine D. How will this difference in radius affect:

(a) the speed, in metres per second, at which the circumference of the drum moves?

(b) the speed at which water flies out of the holes?

Give reasons for your answers.

> ### REMEMBER
>
> To keep an object moving in a circle, a constant centripetal force must act on the object towards the centre of the circle.
>
> The centripetal force needed is greater:
> - the greater the mass of the object
> - the greater the speed of the object
> - the smaller the radius of the circle.

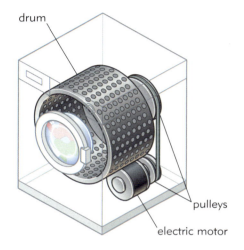

drum

pulleys

electric motor

force of drum on clothes

The drum spins at over 1000 rpm (revolutions per minute)

Where there are holes in the drum, there is no centripetal force from the drum. So water flies out through the holes at a tangent (to the drum).

What you need to remember

Circular motion 3: spin driers

You should be able to interpret information about circular motion in the way that you have on this page.

7

Circular motion 4: cornering on a cycle

When you ride round a corner on a bicycle, you travel in an approximately circular path (see diagram).

This means that there must be a centripetal force to produce the necessary acceleration towards the centre of the circle.

1 (a) What supplies the centripetal force?

   (b) What do you think will happen if this force is not big enough to keep the bicycle moving in a circle at the speed the cycle is travelling?

   (c) Suggest the weather conditions when this is most likely to happen. Explain your answer.

■ How big a centripetal force is needed?

The size of the centripetal force that is needed to keep a bicycle moving in a circle depends on several factors.

2 Explain how each of the following affects the size of the centripetal force needed:

   (a) taking the corner at a lower speed;

   (b) taking the corner more tightly (as shown by the green line on the diagram);

   (c) riding a heavier bicycle.

> **REMEMBER**
>
> To keep an object moving in a circle, a constant centripetal force must accelerate the object towards the centre of the circle.
>
> The centripetal force needed is greater:
> - the greater the mass of the object
> - the greater the speed of the object
> - the smaller the radius of the circle.

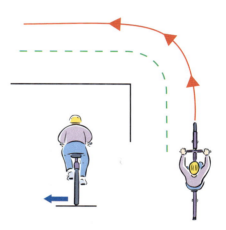

The friction forces of the road on the tyres supply the centripetal force. If friction cannot supply a big enough force, the cycle will slide sideways.

For Higher Tier students only

3 Use information from the diagram to explain why a cycle racing track is banked.

A cycle racing track is steeply banked.

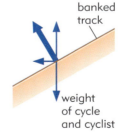

banked track

weight of cycle and cyclist

The force of the banked track on the cycle can be split into:

- an upwards part that balances the weight of the cycle + cyclist

- a sideways part acting towards the centre of the circular track.

What you need to remember

Circular motion 4: cornering on a cycle

You should be able to interpret information about circular motion in the way that you have on this page.

8

Momentum

How hard it is to stop a moving object depends on the mass of the object and on how fast it is moving.

1 Look at the diagrams. Which is it harder to stop moving:

(a) a 30 tonne lorry or a 1 tonne car that is moving at the same speed?

(b) a car that is moving at 10 metres per second or the same car that is moving at 20 metres per second? Give reasons for your answers.

The greater the mass of an object, and the faster it is moving, the harder it is to stop the object.
This is because the object has more **momentum**.

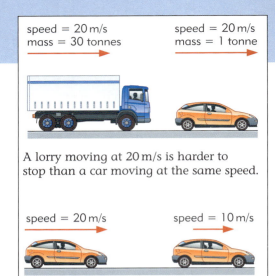

A lorry moving at 20 m/s is harder to stop than a car moving at the same speed.

A car moving at 20 m/s is harder to stop than the same car moving at 10 m/s.

■ Calculating momentum

You can calculate the momentum of a moving object like this:

momentum	=	**mass**	×	**velocity***	[* see Box]
(kilogram metres/second, **kg m/s**)		(kilograms, kg)		(metres/second, m/s)	

Example

A 25 kg supermarket trolley is moving at 2 m/s.
Calculate its momentum.

momentum = mass × velocity [or speed]
      = 25 kg × 2 m/s
      = 50 kg m/s

Speed and velocity

The velocity of an object is its **speed** in a particular direction.

Momentum is mass × velocity.

This means that momentum not only has a size, or **magnitude**, but it also has a **direction**.

Sometimes we only want to know how much momentum a moving object has, so we can use:

magnitude (size) of momentum = mass × speed

2 What is meant by the velocity of an object?

3 How much momentum has the car shown in the diagram?

4 How many times more momentum has a 30 tonne lorry than a 1 tonne car travelling at the same speed?

speed = 30 m/s

mass = 750 kg

What you need to remember [Copy and complete using the **key words**]

Momentum

The greater the mass of an object, and the greater its speed, the more _____ the object has.
The momentum of an object not only has a size (or _____), it also has a _____.
You can calculate momentum like this:

momentum	=	_____	×	_____*
(kilogram metres/second, _____)		(kilograms, kg)		(metres per second, m/s)

[*The velocity of an object is its _____ in a particular direction.]

H3 For Higher Tier students only

Changing momentum

You can change the momentum of a moving object. The diagram shows how you can do this.

1 Copy and complete the sentences.

To change the momentum of a moving object, you must exert a _____ on it.

To increase the momentum of the moving object, the force must act in the same _____ as the object is moving.

To reduce the momentum of an object moving in one direction you must exert a force on the object in the _____ direction.

You can also give a stationary object some momentum by exerting a force on it.

■ Calculating changes in momentum

To produce a bigger change in momentum you can:

■ use a bigger force

■ apply the force for a longer time.

You can calculate the change in momentum produced by applying a force, like this:

change in momentum   = force   × time
(kilogram metres/second, kg m/s)   (newtons, N)   (seconds, s)

2 (a) Calculate the momentum given to the golf ball shown in the diagram.
   [Begin by stating the above formula.]

   (b) Use your answer to (a) to calculate the speed of the ball just after it has been hit.

REMEMBER

momentum   =   mass × velocity
(kilogram metres/second, (kilograms, (metres/second,
kg m/s)   kg)   m/s)

Momentum, like velocity, not only has magnitude (size) but also has direction.

velocity = 25 m/s

A force acting in this direction increases the momentum of the car.

A force acting in this direction reduces the momentum of the car.

golf club

golf ball (mass = 40 g)

movement of club

The golf club is in contact with the ball for $\frac{1}{100}$ of a second. During this time it exerts an average force of 200 N.

Using your knowledge

1 Calculate the average braking force of the car shown in the diagram.
[Start your calculation as shown in the Box below.]

Re-arranging the formula to calculate force

change in momentum = force × time

So:

$$\frac{\text{change in momentum}}{\text{time}} = \text{force}$$

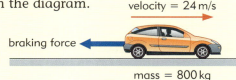

velocity = 24 m/s

braking force

mass = 800 kg

When the driver brakes hard, it takes 5 seconds for the car to stop.

H4 For Higher Tier students only

Explosions

When you fire a rifle, a small explosion sends a bullet flying very quickly out of the barrel.

But the bullet is not the only thing that moves as a result of the explosion (see diagram).

1 Describe the <u>two</u> movements that occur when a rifle is fired.

To understand what happens when a rifle is fired and in other explosions, we need to use the idea of momentum.

■ Momentum in explosions

When an explosion occurs, no <u>external</u> force acts on the object(s) involved in the explosion.

This means that during an explosion the total momentum of the object(s) does not change.

We say that momentum is <u>conserved</u>.

This is possible because momentum has direction as well as size. When things fly apart after an explosion, the momentum in any particular direction of moving parts is always balanced by the momentum in the opposite direction of other moving parts.

2 Look at the diagrams of a rifle before and after it is fired.

(a) What is the <u>total</u> momentum of a rifle <u>and</u> the bullet inside it before the rifle is fired?

(b) What happens to the total momentum of the object(s) involved during any explosion?

(c) What is the <u>total</u> momentum of the rifle <u>and</u> the bullet after the rifle has been fired?

(d) Both the bullet and the rifle <u>do</u> have momentum after the rifle has been fired. How does the magnitude and the direction of the rifle's momentum compare with the momentum of the bullet?

rifle moves to the left   bullet moves to the right

When the rifle is fired, the bullet moves in one direction and the rifle moves in the opposite direction.
We say that the rifle <u>recoils</u>.

REMEMBER

momentum = mass × velocity
(kilogram metres/second,   (kilograms,   (metres/second,
kg m/s)   kg)   m/s)

Momentum, like velocity, not only has magnitude (size) but also has direction.

To change the momentum of an object, a force must act on the object.

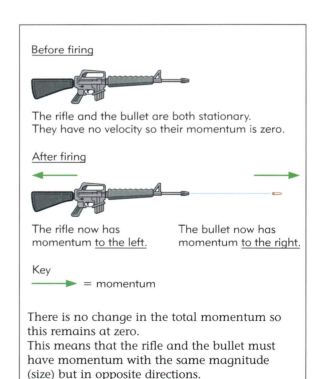

Before firing

The rifle and the bullet are both stationary.
They have no velocity so their momentum is zero.

After firing

The rifle now has   The bullet now has
momentum <u>to the left</u>.   momentum <u>to the right</u>.

Key

——▶ = momentum

There is no change in the total momentum so this remains at zero.
This means that the rifle and the bullet must have momentum with the same magnitude (size) but in opposite directions.

◼ Calculating what happens in an explosion

The first diagram shows:

- ◼ the mass of a rifle
- ◼ the mass of a bullet
- ◼ the speed of the bullet after the rifle has been fired.

You can use this information to calculate the speed of recoil of the rifle.

Before firing the <u>total</u> momentum of the rifle and the bullet is zero.
So the total momentum <u>after</u> firing is also zero.

So ← →
momentum of rifle = momentum of bullet

mass of × speed of = mass of × speed of
  rifle   rifle   bullet   bullet

$$\text{speed of rifle} = \frac{\text{mass of bullet} \times \text{speed of bullet}}{\text{mass of rifle}}$$

$$= \frac{0.01 \times 150}{2}$$

$$= 0.75 \, \text{m/s}$$

3 Use the information on the diagrams to calculate:

(a) the speed of recoil of the field gun

(b) the mass of the hand gun.

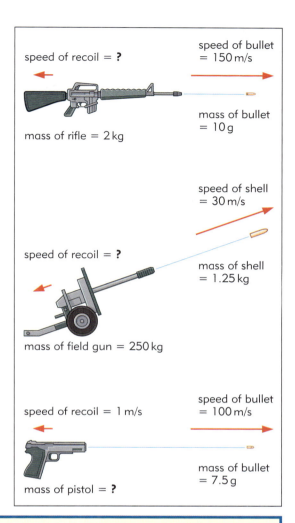

speed of recoil = **?** speed of bullet = 150 m/s

mass of rifle = 2 kg mass of bullet = 10 g

speed of shell = 30 m/s

speed of recoil = **?** mass of shell = 1.25 kg

mass of field gun = 250 kg

speed of recoil = 1 m/s speed of bullet = 100 m/s

mass of pistol = **?** mass of bullet = 7.5 g

Using your knowledge

1 The rocket of a 500 kg spacecraft is given a short 'burn' to increase its speed. Hot gas is ejected with a momentum of 1200 kg m/s.

(a) By how much does the momentum of the spacecraft increase?

(b) By how much does the speed of the spacecraft increase?

(c) If the hot gas is ejected at 80 m/s, what mass of gas is ejected?

2 Atomic bombs, like the ones that were dropped on Hiroshima and Nagasaki in 1945, explode <u>above</u> the ground.

The diagram shows an atomic bomb just before it explodes. What can you say about the total momentum of the fragments of the bomb immediately after it explodes?

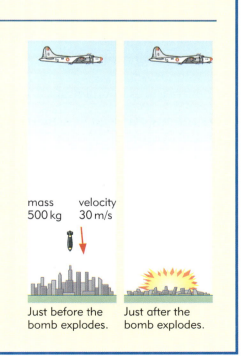

mass velocity
500 kg 30 m/s

Just before the Just after the
bomb explodes. bomb explodes.

H5 For Higher Tier students only

Collisions

When two objects collide, each object exerts a force on the other object (see diagram).

1 Copy and complete the sentences.

When two objects collide, the forces that they exert on each other:

- are the same _____
- act for the same length of _____
- are in opposite _____ .

As a result of the force that acts on it, the momentum of each object changes.

2 When two objects collide, how do their changes in momentum compare:

(a) in magnitude (size)?

(b) in direction?

Give reasons for your answers.

■ Momentum in collisions

When two objects collide, the momentum of each changes by the same amount.
But their changes in momentum are in opposite directions.
So the overall change in momentum during a collision is zero.
In other words, momentum is <u>conserved</u>.

We can use this idea to explain what happens in collisions.

3 Look at the diagrams of two different collisions. In each case:

(a) describe what happens

(b) use the idea of momentum to explain why this happens.

When snooker balls collide they bounce off each other. We say that these collisions are **elastic**.

When blobs of Blu-Tack collide, they stick together. We say that these collisions are **inelastic**.

When two moving objects collide ...

... they exert equal forces on each other, for the same time but in opposite directions.

> ### REMEMBER
>
> | momentum | = | mass | × | velocity |
> |---|---|---|---|---|
> | (kilogram metres/second, kg m/s) | | (kilograms, kg) | | (metres/second, m/s) |
>
> Momentum, like velocity, not only has magnitude (size) but also has direction.
>
> To change the momentum of an object, a force must act on the object.
>
> change in momentum = force × time

Colliding snooker balls

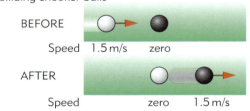

Momentum is conserved.
So the collision must transfer all the momentum of the left-hand ball to the right-hand ball.
The balls have the same mass so the final velocity (and speed) of the right-hand ball is the same as that of the left-hand ball at the start.

Colliding blobs of Blu-Tack (of equal mass)

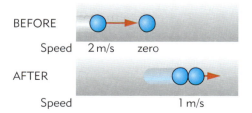

Momentum is conserved.
When the blobs stick together, the mass is doubled.
So the final velocity (and the speed) of the combined blob is half that of the left-hand blob at the start.

Kinetic energy in collisions

The kinetic energy of a moving object also depends on the mass and the velocity of the object, but in a different way from the momentum (see Box).

Another difference between kinetic energy and momentum is that kinetic energy is <u>not</u> always conserved during collisions.

4 Using the information from the diagrams on page 62, calculate the overall change in kinetic energy (if any):

  (a) in the elastic collision between the two snooker balls, each with a mass of 120 g

  (b) in the inelastic collision between the two blobs of Blu-Tack, each with a mass of 10 g.

5 Copy and complete the sentences.

In an inelastic collision, there is less _____ energy after the collision than there was before the collision.
In an _____ collision, there is no loss of kinetic energy.

Most collisions, for example collisions between motor vehicles, are inelastic. The collisions between snooker balls are <u>almost</u> elastic, but they are not <u>completely</u> elastic. The collisions between molecules in gases are <u>perfectly</u> elastic.

Momentum calculations in collisions

The Box shows how you can use the idea of momentum to calculate what happens in a collision.

6 Calculate what happens if the lorry's speed is 30 m/s and the car is moving at 20 m/s in the same direction as the lorry.

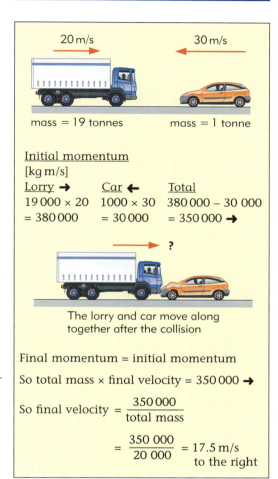

20 m/s   30 m/s

mass = 19 tonnes   mass = 1 tonne

Initial momentum
[kg m/s]

Lorry ➡	Car ⬅	Total
19 000 × 20	1000 × 30	380 000 – 30 000
= 380 000	= 30 000	= 350 000 ➡

?

The lorry and car move along together after the collision

Final momentum = initial momentum

So total mass × final velocity = 350 000 ➡

So final velocity = $\dfrac{350\,000}{\text{total mass}}$

$= \dfrac{350\,000}{20\,000} = 17.5$ m/s to the right

Using your knowledge

1 For the collision shown in the Box:

  (a) calculate the total kinetic energy of the lorry and the car before and after the collision

  (b) use your answer to (a) to show that the collision is inelastic

  (c) suggest what happens to the lost kinetic energy.

2 Two objects with the same mass (*m*) are travelling towards each other with the same speed (*v*).

Describe, and then explain, what happens:

(a) if the collision is perfectly elastic

(b) if the collision is completely inelastic

(c) if the collision is almost elastic.

The Earth's structure

The Earth's surface is slowly changing all the time. Mountains are gradually being worn away and huge forces are pushing up rocks to make new mountains.

To understand the forces that make mountains, we need to know what scientists have discovered about the Earth's structure.

1 What do we know about the <u>outside</u> of the Earth?

▌Inside the Earth

Even the deepest drill cannot make a hole through the Earth's **crust**. However, scientists think that the Earth is made of different layers. They have found out a bit about these layers by studying the way vibrations from explosions and earthquakes travel through the Earth. The diagram shows what these layers are probably like.

2 Imagine that you could drill a hole through the Earth to the centre.

Copy and complete the following sentences to say what you would find on the way through.

(a) First the drill would go through the solid rock in the Earth's _____.

(b) Next, the drill would reach the denser rocks of the _____. This has several layers.

(c) About halfway through to the centre of the Earth the drill would reach the outer _____. This is made of liquid _____ and _____ metals.

(d) The inner core is made of the same two metals but they are _____.

Sometimes part of the lower crust or the outer mantle melts. We call the molten rock magma.

The Earth has a much greater mass than if it were only rock inside. So it has a higher average **density** than rock.

3 Why is the average density of the Earth much greater than the rocks that we find in the Earth's crust?

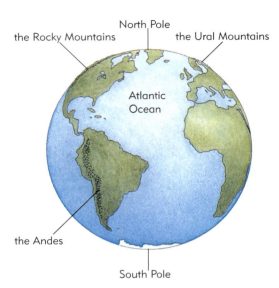

The Earth is a **sphere** that is slightly flattened at the poles. Though there are high mountains and deep oceans, these are <u>very small</u> compared with the size of the Earth itself. If the Earth were the size of a snooker ball it would be just as smooth.

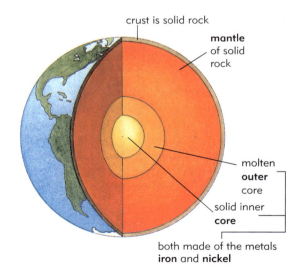

Nickel and iron are much **denser** materials than rock. (In other words, nickel and iron have a much greater mass than the same volume of rock.) The mantle is made of denser rock than the crust. This rock is solid, but the upper part is so near to its melting point that it is hot enough to **flow** very slowly.

■ Changes to the Earth's crust

The Earth's crust is changing all the time. Some of the changes happen quickly, but other changes are slow and can take millions of years. Look at the two diagrams and the two text boxes.

4 Write down <u>two</u> things that can change the Earth's crust quickly.

New rocks can form when things happen to the Earth's crust.

5 Copy the headings and complete the table.

What happens to the Earth's crust	Kind of rock formed

Some of the changes that happen to the Earth's crust wear away the hills and mountains.
Other changes help to form new hills and mountains.

6 Write down <u>two</u> changes that help to make new mountains.

When new mountains are made, rocks can be pushed up thousands of metres. This needs huge forces.

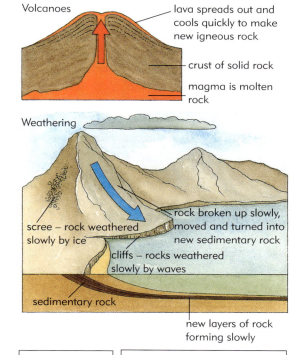

Volcanoes — lava spreads out and cools quickly to make new igneous rock — crust of solid rock — magma is molten rock

Weathering — scree – rock weathered slowly by ice — rock broken up slowly, moved and turned into new sedimentary rock — cliffs – rocks weathered slowly by waves — sedimentary rock — new layers of rock forming slowly

Earthquakes
The crust can suddenly move or break in an earthquake.

Rocks folding
Large forces can make rocks fold slowly. When rocks are folded upwards, they can form new hills and mountains. Some rocks can be changed into metamorphic rocks.

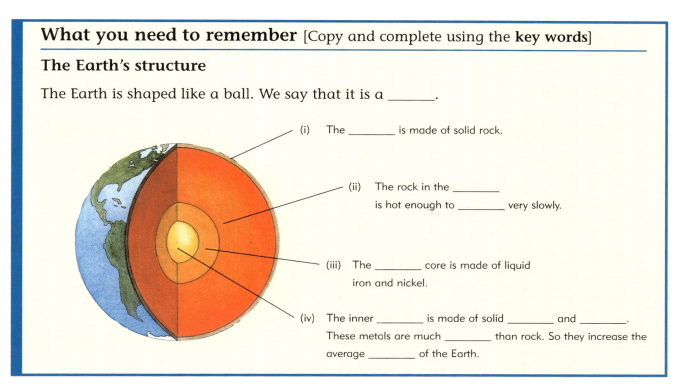

What you need to remember [Copy and complete using the **key words**]

The Earth's structure

The Earth is shaped like a ball. We say that it is a _____.

(i)   The _____ is made of solid rock.

(ii)   The rock in the _____ is hot enough to _____ very slowly.

(iii)   The _____ core is made of liquid iron and nickel.

(iv)   The inner _____ is made of solid _____ and _____. These metals are much _____ than rock. So they increase the average _____ of the Earth.

10

Movements that make mountains

Movements of the **Earth's crust** can push up rocks for thousands of metres. We find rocks at the tops of mountains that formed from sediments in the sea. It takes a long time and huge forces for this to happen.

1 Look at the diagram. How do we know that rocks can be pushed up thousands of metres?

This fossil came from an animal that lived in the sea millions of years ago.

How mountain ranges are formed

Different ranges of mountains were pushed up at different times during the Earth's history. As old mountains wore down in one place, new mountains were pushed up in others. The Scottish Highlands were part of a huge range of high mountains. They formed 450 to 550 million years ago. Since then, weathering and **erosion** have worn them down. Higher mountains such as the Alps are much younger. They formed 7 to 25 million years ago. The Andes and the Himalayas are even higher and they are still being pushed up.

2 Younger mountain ranges are usually higher than older mountain ranges. Explain why.

The movements in the Earth's crust during mountain-building produce high **temperatures** and **pressures**. This alters the structure of rocks. So we find **metamorphic** (changed) rocks in places where mountains are forming. We also find metamorphic rocks where mountains formed in the past.

3 Why does mountain-building also result in the formation of metamorphic rocks?

mountain ranges produced by rock being pushed upwards

Volcanoes

Relatively small mountains called volcanoes are formed by rock movement of a completely different kind. The diagrams show what happens.

4 Copy and complete the sentences.

Molten rock is forced up through cracks in the Earth's crust and flows out as _____.

This _____ as it cools and gradually builds up a _____ -shaped mountain.

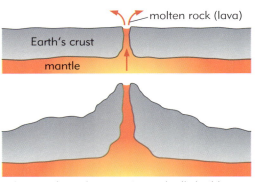

A cone-shaped mountain gradually builds up

Volcanic eruptions

When a volcano erupts, it can be very dangerous for the people who live nearby.

The photos show three different types of **volcanic** eruption.

Hot molten lava flowing slowly from a volcano can destroy houses.
But people have time to escape. Sometimes the lava flow can be diverted away from villages.

5 Which <u>two</u> types of eruption are particularly dangerous for the people who live nearby?

Predicting when a volcano will erupt

Because they can be so dangerous, it is important to be able to **predict** when a volcano is going to erupt.

Scientists measure temperatures, pressures and the gases given off. This can be difficult and dangerous work. They can sometimes measure an increase in pressure and say that there will be an eruption in the next few months. But there are so many factors involved that, often, they cannot be more accurate.

6 Use information from this page to write a short article about 'Predicting volcanic eruptions'.

During a pyroclastic flow, hot ash and gases roll down the sides of a volcano at speeds of up to 200 kilometres an hour.

In 1902, Mont Pelee in Martinique erupted unexpectedly. It took just 1 minute for the town of St. Pierre to be completely covered by a pyroclastic flow. 29 000 people were killed.

The pressure below a volcano can build up so much that it blows off millions of tonnes of rock from the top or side of the volcano in a huge explosion. This happened to Mount St. Helens, in the USA, in 1980.

Scientists predicted the eruption and people were moved away from the upper slopes. But the scientists were taken by surprise when the eruption caused all this damage.

What you need to remember [Copy and complete using the **key words**]

Movements that make mountains

Weathering and _____ wear mountains away.
Large-scale movements of the _____ _____ over millions of years cause new mountains to form.
Mountain-building involves high _____ and _____ .
So _____ rocks form at the same time as new mountain belts.
Smaller mountains can be built by _____ activity.
Volcanic eruptions can be very dangerous, but it is difficult to _____ when they will happen.

11 How does the Earth's crust move?

Large mountain ranges are formed when rocks of the Earth's crust are slowly forced upwards.

This happens because of movements of the Earth's crust and the layer of the mantle just below the crust.

The Earth's crust and the upper part of the mantle are called the **lithosphere**.

The lithosphere is not made of one big piece of rock. Cracks split it into very large pieces called **tectonic plates**. The map shows some of them.

The plates **move** all the time. They do not move very fast, just a few **centimetres** (cm) each year. But these small movements add up to big movements over a long time.

1 Copy and complete the following sentences.

   Britain is on the _____ plate. North and South America are on the American _____.

2 (a) Which way is the American plate moving?

   (b) Which way is the Eurasian plate moving?

   (c) What is happening to the distance between America and Europe?

3 A plate moves about 5 cm each year.
   How far will the plate move in 1000 years?

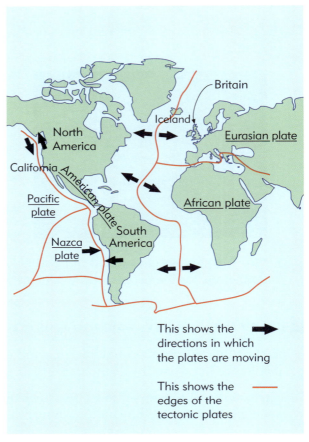

This shows the directions in which the plates are moving ➡

This shows the edges of the tectonic plates ⎯

How some tectonic plates are moving.

■ How do we know where the edges of plates are?

The lithosphere is unstable in the places where two plates meet. So these are the places where we would expect **earthquakes** and **volcanic** eruptions.

4 (a) Where in North America would you expect frequent earthquakes?

   (b) Where in northern Europe, would you expect a lot of volcanic activity?

5 Why do we have few earthquakes in Britain?

How fast do tectonic plates move?

Tectonic plates move hardly any faster than your fingernails grow. They move at a slower rate than your hair grows.

You can estimate these rates as follows.

- Decide how many times a year your hair (or fingernails) is cut.
- Decide how many centimetres (or millimetres) are removed each time.
- Use these figures to work out the annual rate of growth.

The **shapes** of South America and Africa fit together.

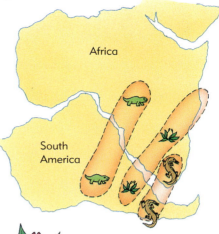

How do we know that plates move?

South America and Africa are on different plates. These plates have been moving away from each other for millions of years. Long ago, South America and Africa must have been together. Look at the diagrams.

The Earth today

The Earth millions of years ago

6 The shapes of South America and West Africa tell us that they were once together. Explain why.

7 The rocks in South America and Africa also tell us that they were once joined together. Explain why.

Fossils of this fern have been found all over Africa and South America.

Fossils of this reptile have been found in Brazil and in Africa.

Fossils of this reptile have been found in Argentina and in southern Africa.

How moving plates can make new mountains

Some of the plates of the Earth's crust are moving away from each other. Other plates are pushing into each other. When two plates move towards each other they can force rocks upwards to give new **mountains**.

8 Look at the diagram.

(a) Which <u>two</u> plates are pushing into each other?

(b) Which mountains have been formed by these

What you need to remember [Copy and complete using the **key words**]

How does the Earth's crust move?

The Earth's _____ is cracked into large pieces. We call these _____ _____.

The plates _____ very slowly, just a few _____ each year.

Millions of years ago, South America and Africa were next to each other.

We know this because:

■ their _____ fit together well

■ they have rocks containing the same _____.

In some places, tectonic plates push together.

This forces some rocks upwards and makes new _____.

The boundaries between plates are where _____ and _____ eruptions mainly occur.

What keeps the Earth's crust moving?

The Earth's **mantle** is solid. But it is so close to its melting point that it can flow like a liquid. It flows very slowly. When there are movements in the mantle, the tectonic plates above also move.

> **REMEMBER**
>
> The Earth's lithosphere is cracked into pieces called tectonic plates. These plates move around slowly.

What makes liquids move?

If a liquid gets hot, it moves around.
The diagrams show what happens.

1 Copy and complete the following sentences.

Water _____ around when you heat it. This is because hot water _____ and cold water moves _____ to take its place.

These movements are called **convection** currents.

> **For Higher Tier students only**
>
> **Ideas you need from**
> **Waves and Radiation**
>
> Scientists know that the Earth's mantle is solid because longitudinal S-waves produced by earthquakes can travel through it.
>
> These waves cannot travel through the liquid outer core of the Earth.

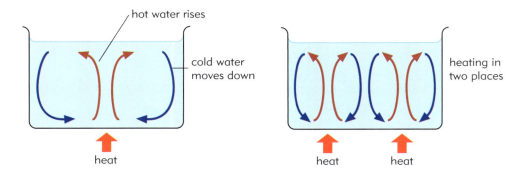

Convection currents inside the Earth

Heat produced inside the Earth causes slow convection currents in the mantle.

Look at the diagram.

2 Copy and complete the following sentences.

Convection currents in the mantle:

■ make plates A and B move _____

■ make plates B and C move _____.

The plates can move because they 'float' on top of the _____. There are very slow _____ currents in the mantle.

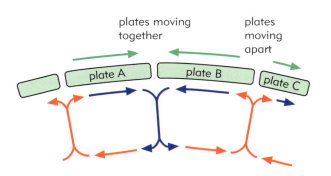

Movements of the mantle move the tectonic plates.

How does the inside of the Earth keep hot?

Something must be heating up the mantle or there wouldn't be any convection currents.

Radioactive substances inside the Earth produce the heat that is needed. The diagram shows how they do this.

3 Copy and complete the following sentences.

Uranium atoms _____ up into smaller atoms of _____ and _____. This change also releases some _____.

4 Radioactive substances will keep on heating up the inside of the Earth for a long time to come. Explain why.

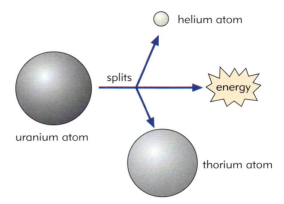

The Earth formed about 4.5 billion (4 500 000 000) years ago. Since then, about half of the uranium atoms have split up.

What happens when the tectonic plates move apart?

When tectonic plates push against each other, new mountains form.

The diagram shows what happens when plates move apart.

5 (a) What type of new rock spreads out through the cracks between plates?

(b) Why is this type of rock formed?

6 Look at the map. Write down the name of a country where basalt is forming.

The edges of the plates that are moving apart are usually under the sea.

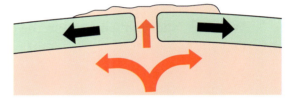

When two plates move apart, magma fills the gap. The magma quickly cools to form the igneous rock called basalt.

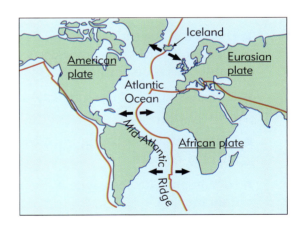

What you need to remember [Copy and complete using the **key words**]

What keeps the Earth's crust moving?

Tectonic plates move because of _____ currents in the _____ below the Earth's lithosphere.

The energy that produces the currents comes from _____ substances inside the Earth.

13

Earthquakes

Earthquakes are caused by the movement of **tectonic plates**. Most earthquakes occur at the boundaries between tectonic plates.

Earthquakes mainly happen where two plates are sliding along each other.

1 How is the tectonic plate movement that causes earthquakes different from the tectonic plate movement that produces new mountain chains?

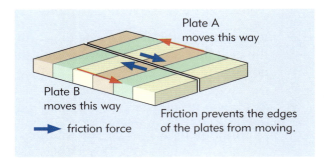

Plate A moves this way

Plate B moves this way

➡ friction force

Friction prevents the edges of the plates from moving.

■ How do sliding plates cause earthquakes?

The diagrams show what happens when two tectonic plates slide along each other.

2 Write the following sentences in the correct order to explain how an earthquake occurs.

- Eventually, the friction force between the two plates cannot increase any further.

- The main parts of two adjacent plates move in opposite directions but do not move apart.

- The friction forces between the plates cause the rocks at the edge of each plate to bend.

- The rock at the edge of each plate straightens out. The energy released sends shock waves through the plates. This is an earthquake.

- The edges of the plates suddenly slide along each other.

- The more the rock bends, the greater the stress forces in the rock.

- Friction prevents the edges of the plates from moving.

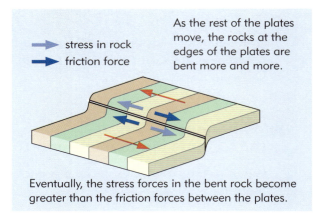

➡ stress in rock
➡ friction force

As the rest of the plates move, the rocks at the edges of the plates are bent more and more.

Eventually, the stress forces in the bent rock become greater than the friction forces between the plates.

So the edges of the plates suddenly slide, releasing all the energy stored up in the bent rock.

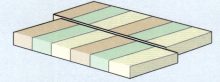

Shock waves are sent through the plates. This is an earthquake.

■ How often do earthquakes occur?

Small earthquakes happen very frequently but severe earthquakes happen far less often. Every year, for example, there are about 100 earthquakes that damage buildings. But there is an earthquake that totally destroys most buildings only every 5–10 years.

When the tectonic plates on either side of the San Andreas fault line slip along each other they cause an earthquake.

Can earthquakes be predicted?

Because of the way they are caused, severe earthquakes in a particular place tend to occur at approximately regular intervals. But these intervals are often tens, or even hundreds, of years long. This means that scientists can only **predict** the next earthquake to within a few years. This is no use at all for deciding when to evacuate a town to save lives.

Sometimes there are lots of very tiny earthquakes a short time before a major earthquake. These can be detected using seismographs. The ground may also start to tilt and the depth of water in wells may suddenly change just before an earthquake happens. These warning signs do not, however, always appear.

3 (a) How can scientists predict earthquakes?

(b) Why are their predictions not usually very useful?

Movement of rocks along the San Andreas fault caused the earthquake that destroyed this bridge in California.

50 000 buildings in Kobe, Japan, were destroyed in the 1995 earthquake. 5000 people were killed.

How can the deaths and damage caused by earthquakes be limited?

Even when there is a warning of a serious earthquake, the amount of warning is often much shorter than it is for volcanic eruptions, e.g. minutes rather than hours.

Since there is often not time to evacuate buildings before an earthquake, it is important that buildings are designed to withstand all but the most severe earthquakes. Since 1980, in Japan, all new buildings have had to meet strict regulations to make them safer in earthquakes.

4 What can people in older buildings do to protect themselves against earthquakes?

Japanese earthquake advice

The islands of Japan lie along the boundary between the Philippine plate and the Pacific plate.
Earthquakes are common.

People living or working in pre-1980 buildings are advised:

- to install extra supports so that the buildings don't collapse so easily
- to bolt gas appliances to the foundations and to connect them with flexible pipes
- to fix shelves securely to walls and to put heavy objects on the lower shelves
- not to place beds near to windows.

What you need to remember [Copy and complete using the **key words**]

Earthquakes

Earthquakes tend to happen where the boundaries between _____ _____ slide along each other.

Scientists often know <u>approximately</u> when large earthquakes will happen, but they cannot usually _____ them accurately enough, or early enough, for the information to be very much use.

[You should be able to interpret information about predicting earthquakes just like you have on these pages.]

14 Changing ideas about the Earth

Until about 200 years ago, most people believed that the mountains, valleys and seas on the Earth had always been just like they are today. Many of these people thought the Earth was created only a few thousand years ago.

Then geologists started to study the rocks and to think about how they were formed. They realised that the Earth must be many millions of years old.

1 Why did they think that the Earth must be millions of years old?

2 What else did they then need to explain?

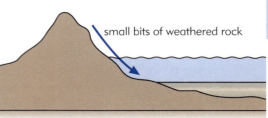

Thick layers of sedimentary rock must take millions of years to form.

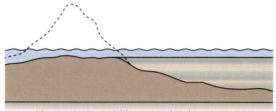

Mountains would be completely worn away over millions of years. So geologists need to explain how new mountains are formed.

◼ A cooling, shrinking Earth

The diagrams show one theory about how new mountains are formed.

3 Write down the following sentences in the correct order.

◼ The molten core carries on cooling, but more and more slowly. It shrinks as it cools.

◼ The Earth began as a ball of hot, molten rock.

◼ The shrinking core makes the crust wrinkle. The high places become mountains, the low places become seas.

◼ As the molten rock cooled, a solid crust formed.

According to this theory, the Earth can't be more than about 400 million years old or it would be cool and completely solid by now.

◼ Problems for the shrinking Earth theory

We now know that the Earth is a lot older than 400 million years. We know this because the Earth contains quite a lot of radioactive elements such as uranium. The atoms of these elements gradually decay (break up). Heat is released as they do so.

4 What effect does this have on the Earth's core?

5 The oldest rocks on Earth are more than 3.5 billion years old. How do scientists know this?

What scientists used to think

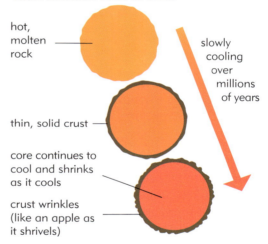

What they think now

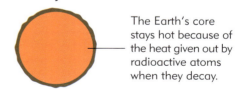

The Earth's core stays hot because of the heat given out by radioactive atoms when they decay.

Dating rocks

Scientists can measure:

◼ the amounts of radioactive atoms in rocks

◼ the amounts of atoms produced when the radioactive atoms decay.

This tells them how old the rocks are.

■ The idea of a moving crust

Scientists now think that mountains are formed by the Earth's crust moving about. Alfred Wegener first suggested this idea during the early years of the twentieth century. But most scientists didn't agree for another 50 years. This idea was called the theory of continental drift.

6 Why was Wegener's theory of crustal movement called continental drift?

7 What evidence did scientists have for continental drift?

8 Why did many scientists not agree?

During the 1950s, scientists started to explore the rocks at the bottom of the oceans. The diagrams show what they found and how they explained it.

9 Copy and complete the following sentences.

Under the oceans are long _____ ridges.

These are made of rock that is quite _____.

The sea floor under the ocean is moving _____.

Magma from below the Earth's _____ moves up to make new rock.

The new evidence convinced scientists that the Earth's crust is made of a small number of separate sections called plates. Under the oceans these plates are moving apart. But in some places these plates are moving towards each other. This pushes rock upwards to make new mountains.

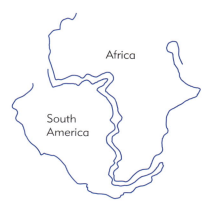

Some scientists suggested that South America and Africa must once have been together. Other scientists said that there was no way they could possibly have moved apart.

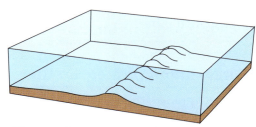

There are long mountain ridges underneath the ocean. They are made of young rocks.

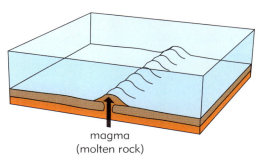

magma
(molten rock)

Sections of crust on the sea floor are moving apart. New rock forms to fill the gap.

What you need to remember

Changing ideas about the Earth

You should be able to:

- describe the 'shrinking Earth' model of how mountains are formed
- explain why this model has been replaced by the idea of the Earth's crust being made up of moving plates
- explain why the idea of moving continents was not accepted by most scientists until about 50 years after Wegener suggested it.

H6 For Higher Tier students only

More about tectonic plates

The Earth's lithosphere is cracked into a number of **tectonic plates**, which are slowly moving. Movement is caused by **convection currents** in the Earth's mantle. Tectonic plates vary in thickness from 50 to 200 km.

The boundaries between the plates are called plate margins. At these plate margins, three different things can happen:

- the plates can slide past each other
- the plates can move towards each other
- the plates can move apart.

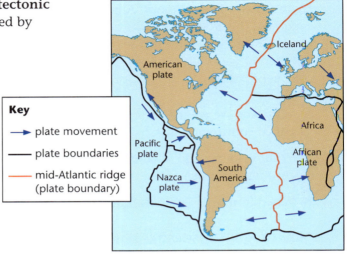

1 Where on the map are the plates sliding past each other?

■ Plates sliding past each other

When two plates slide past each other, no crust is created or destroyed, but friction between the two plates leads to earthquakes.

San Francisco, in California, USA, sits on top of the San Andreas fault, where the movement is about 5 cm per year. Small earthquakes happen often but Californians fear 'the big one'.

2 (a) What do we call the margin between two plates that slide past each other?

   (b) What effect do the sliding plates have?

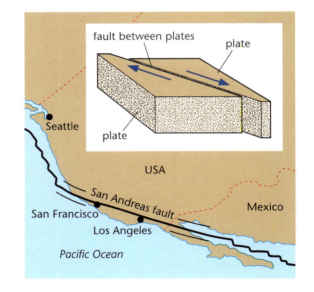

■ Plates moving towards each other

Oceanic crust is denser than continental crust. So when they are pushed against each other the oceanic crust is forced down or **subducted**. The huge forces crumple and push up the layers of rock in the continental crust to form a fold mountain chain. In places where the subducted crust melts, the magma rises and volcanoes form.

3 (a) Outline what happens when an oceanic and a continental plate move towards each other.

   (b) Where, on the map at the top of the page, are mountains being made in this way?

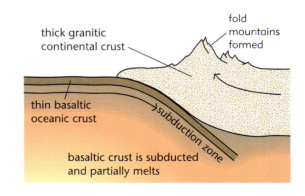

Plates moving apart

Where two plates are moving apart, the oceanic crust cracks. Magma rises to fill the cracks. It cools and solidifies forming new oceanic crust. New crust is added all the time as new cracks open.
This is **sea floor spreading**.

Sea floor spreading is happening in the middle of the Atlantic Ocean in a line between Europe and Africa, and the Americas. We call it the mid-Atlantic ridge.

Occasionally the oceanic ridge rises above sea level to form volcanic islands. For example, the island of Surtsey rose out of the Atlantic Ocean in 1963 from a large submarine volcanic eruption.

4 Explain the positions of the rocks on Iceland and describe what is happening at this plate boundary.

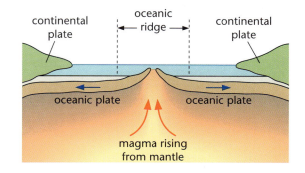

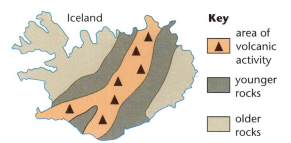

The island of Iceland sits astride an oceanic ridge called the mid-Atlantic ridge.

Scientists have found some interesting support for the theory of sea floor spreading. The Earth's magnetic field reverses from time to time. When magma solidifies, the iron-rich minerals in it line up with the magnetic field at the time.

Scientists looked at the magnetism of the rocks on either side of oceanic ridges. They found a pattern of magnetic stripes parallel to the ridges. This magnetic reversal pattern was the same on both sides of the ridges. In sea floor spreading, new rock moves away on both sides of a ridge, so the pattern was fairly symmetrical.

5 Describe the pattern of magnetic stripes found in rocks in certain places on the sea bed and explain how the stripes were formed.

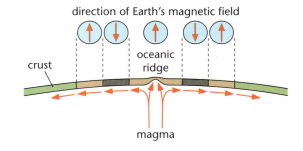

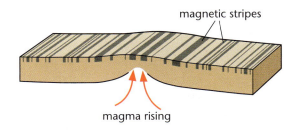

Using your knowledge

1 Explain why marine fossils are found high in the Andes mountains, on the west coast of South America.

2 Africa and South America are now about 5000 km apart. Assuming that plates move apart at an average speed of 2.5 cm per year, how long ago did they form one land mass?

Physics in action

1 Controlling currents in circuits

For an electric current to flow from a battery or power supply you need a **complete** circuit of **conductors**. You can switch off a current by **breaking** the circuit. You can reduce the current flowing through a circuit by increasing the **resistance** of the circuit. You can do this using a **fixed** resistor

or a **variable** resistor.

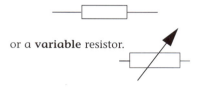

2 Controlling electrical appliances automatically: the thermostat

For some purposes, hand-operated (**manual**) controls are very convenient. Sometimes, however, we prefer automatic control systems, for example temperature controlled by a **thermostat**.
Nowadays automatic control systems are often **electronic**.

3 Electronic control systems 1: the basic idea

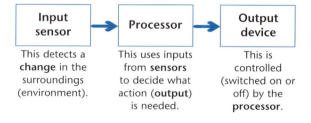

Input sensor	Processor	Output device
This detects a **change** in the surroundings (environment).	This uses inputs from **sensors** to decide what action (**output**) is needed.	This is controlled (switched on or off) by the **processor**.

Component	Symbol
LDR	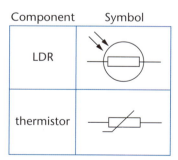
thermistor	

To detect changes in light you can use an **LDR** as an input sensor.
To detect changes in temperature you can use a **thermistor** as an input sensor.

4 Electronic control systems 2: extending the range

Input sensor	What it responds to
push switch	applied pressure
LDR	light
thermistor	temperature
magnetic switch	presence of magnetic field
tilt switch	being tilted

Output device	Transfers energy to surroundings as …
LED	light
motor	movement (kinetic energy)
buzzer	sound
lamp	light
heater	heat (thermal energy)

(*Note that all these output devices also transfer some energy as heat and/or light, in addition to the useful energy transfers listed in the table.*)

The symbol for an LED is:

5 Looking at processors

The processors in electronic control systems are usually made from **logic gates**.

Name of gate	Symbol	For the output of the gate to be on …
OR		The 1st input must be on **or** the 2nd input must be on [or <u>both</u> inputs must be on]
NOT		the input must **not** be on
AND		the 1st input must be on **and** the 2nd input must be on

6 Using logic gates in electronic control systems

[You must be able, when you are presented with a block diagram of a simple electronic system, to describe:

■ what each part of the system does
■ what the whole system does
just as you did on pages 16 and 17.]

7 Another way of looking at processor inputs and outputs

[You need to be able to make, and interpret, truth tables in the way that you did on pages 18 and 19.]

8 How to switch on large currents

Electronic control circuits can usually supply only a small **current**. To control output devices that need a large current, a **relay** is used as a buffer. This device works because it uses a **small** current to switch on a **large** current. Relays can also be used to allow **low** voltage electronic circuits to control appliances that work from the 230 volt **mains** supply.

The symbol for a relay is:

9 How to connect input sensors 1: switches

When a switch is used as an input sensor, it is placed between the **positive** side of a low voltage supply and an **input** to a logic gate. When the switch is on, the input to the logic gate is then on (or 1, or **high**).

10 How to split voltages: the potential divider

The diagram shows a **potential divider** circuit. Components R_1 and R_2 are **resistors**. The voltage supplied (V_{in}) is **shared** between R_1 and R_2.
[You should be able to use the formula

$$V_{out} = V_{in} \times \frac{R_2}{R_1 + R_2}$$

for the potential divider.]

11 How to connect input sensors 2: LDRs and thermistors

To make input sensor circuits, LDRs and thermistors are used as part of a **potential divider**.
The output from the sensor circuit (V_{out}) is used as an input to a **processor**. Changes in the surroundings change the **resistance** of LDRs and thermistors.
This changes the size of V_{out}. The input to the processor may then change from **off** to **on** (or vice versa).

12 Making a moisture sensor

[You should be able to explain, step by step, how an electronic control system works, just as you did on page 26.]

13 How to adjust input sensor circuits

To make a more convenient and more flexible input sensor circuit, you can use a thermistor (or an LDR) and a **variable** resistor in a potential divider circuit. You can then **adjust** the circuit so that V_{out} changes from low to high (or vice versa) at whatever **temperature** (or light level) you choose.

14 What's the delay?

When a capacitor is connected to a battery, electrical **charge** flows into the capacitor.
As the capacitor **charges**, the voltage across the capacitor **increases**.
When a conductor is connected across a charged capacitor, electrical charge flows out of the capacitor and **discharges** it. As this happens, the **voltage** across the capacitor falls.
To make the voltage rise (or fall) more slowly as a capacitor charges (or discharges), you can use a **capacitor** with a greater value or you can connect a higher **resistance** in series.
Capacitors can be used to produce a time **delay** in electronic control circuits.

15 Two types of lens

When parallel rays of light pass through a **converging** lens they come together at a point. When parallel rays of light pass through a **diverging** lens they spread out as if they had all come from a point. In both cases we call this point the **focus** of the lens.
[You should be able to draw what happens to parallel rays of light when they pass through both types of lens.]

16 Two types of image

You can use a converging lens to form an **image** of an object on a screen.
An image that you can form on a screen is called a **real** image.
An image that you cannot form on a screen is called a **virtual** image.
In a camera, a converging lens is used to form an image on a **film**.
The image formed by a camera is:

■ smaller than the **object**
■ **nearer** to the lens than the object.

Forces and motion

1 Turning forces

A force that is applied to an object at a distance from a pivot produces a **turning** effect.
This is called the **moment** of the force and can be calculated using:

moment = **force** × **perpendicular** distance
[newton [newtons, N] between line of action
metres, of force and **pivot**
N m] [metres, m]

2 When the swinging stops

The point on a body where you can think of all of its weight acting is called the **centre of mass**.
When an object is suspended, it comes to rest with the centre of mass vertically below the point of **suspension**. This means that the **weight** of the object does not produce a **turning** effect.
The centre of mass of a symmetrical flat object must lie on each **axis of symmetry**.
[You should be able to describe how to find the centre of mass of a thin sheet of any shape.]

3 How to accelerate without going faster

When an object moves at a steady speed in a circle, its **velocity** is constantly changing.
This means that the object is **accelerating** all the time.
To produce this constant acceleration, a **force** is needed, acting towards the centre of the circle.
This is called a **centripetal** force and it needs to be greater:

■ the greater the **mass** of the object
■ the higher the **speed** of the object
■ the smaller the **radius** of the circle.

4 Circular motion 1: in the solar system and beyond

Astronomical objects often move in approximately **circular** orbits.
For example, the **planets** move around the Sun and the **Moon** moves around the Earth.
To keep an object moving in a circle, a **centripetal** force is needed.
For astronomical objects this force is supplied by **gravity**.
[You should be able to interpret information about circular motion in the way that you did on page 54.]

5 Circular motion 2: inside atoms

Electrons move in approximately **circular** orbitals around the nucleus of an atom.
To keep an object moving in a circle, a **centripetal** force is needed.
For the electrons in an atom, this force is supplied by the **electrostatic** force of attraction between **positive** and **negative** charges.
[You should be able to interpret information about circular motion in the way that you did on page 55.]

6 Circular motion 3: spin driers

[You should be able to interpret information about circular motion in the way that you did on page 56.]

7 Circular motion 4: cornering on a cycle

[You should be able to interpret information about circular motion in the way that you did on page 57.]

8 Momentum

The greater the mass of an object, and the greater its speed, the more **momentum** the object has.
The momentum of an object not only has a size (or **magnitude**), it also has a **direction**.
You can calculate momentum like this:

momentum = **mass** × **velocity***
(kilogram metres/second, (kilograms, (metres per second,
kg m/s) kg) m/s)

[*The velocity of an object is its **speed** in a particular direction.]

9　The Earth's structure

The Earth is shaped like a ball. We say that it is a **sphere**.

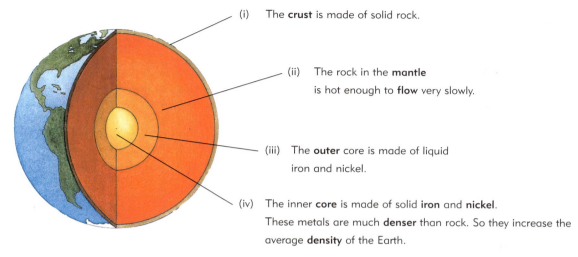

(i)　The **crust** is made of solid rock.

(ii)　The rock in the **mantle** is hot enough to **flow** very slowly.

(iii)　The **outer** core is made of liquid iron and nickel.

(iv)　The inner **core** is made of solid **iron** and **nickel**. These metals are much **denser** than rock. So they increase the average **density** of the Earth.

10　Movements that make mountains

Weathering and **erosion** wear mountains away. Large-scale movements of the **Earth's crust** over millions of years cause new mountains to form. Mountain-building involves high **temperatures** and **pressures**.
So **metamorphic** rocks form at the same time as new mountain belts.
Smaller mountains can be built by **volcanic** activity. Volcanic eruptions can be very dangerous, but it is difficult to **predict** when they will happen.

11　How does the Earth's crust move?

The Earth's **lithosphere** is cracked into large pieces. We call these **tectonic plates**.
The plates **move** very slowly, just a few **centimetres** each year.
Millions of years ago, South America and Africa were next to each other.

We know this because:
■ their **shapes** fit together well
■ they have rocks containing the same **fossils**.

In some places, tectonic plates push together. This forces rocks upwards and makes new **mountains**. The boundaries between plates are where **earthquakes** and **volcanic** eruptions mainly occur.

12　What keeps the Earth's crust moving?

Tectonic plates move because of **convection** currents in the **mantle** below the Earth's lithosphere. The energy that produces the currents comes from **radioactive** substances inside the Earth.

13　Earthquakes

Earthquakes tend to happen where the boundaries between **tectonic plates** slide along each other. Scientists often know <u>approximately</u> when large earthquakes will happen, but they cannot usually **predict** them accurately enough, or early enough, for the information to be very much use.
[You should be able to interpret information about predicting earthquakes just like you did on pages 72 and 73.]

14　Changing ideas about the Earth

You should be able to:
■ describe the 'shrinking Earth' model of how mountains are formed
■ explain why this model has been replaced by the idea of the Earth's crust being made up of moving plates
■ explain why the idea of moving continents was not accepted by most scientists until about 50 years after Wegener suggested it.

Glossary/index

[Note: Concepts that are also to be found in the basic *Science Foundations: Physics (New Edition)* text have not been included unless those concepts are developed further in this text.]

A

acceleration: the rate at which the *velocity* of an object changes 52–53

AND gate: a *logic gate* whose output is on/high/1 when input A <u>and</u> input B are both on/high/1; can be used in the *processor* circuit of an *electronic system* 14, 18

axis of symmetry: a line dividing a flat (2-dimensional) object into two mirror-image halves; the *centre of mass* of the object lies along this line 47

C

capacitor: a device that can store and release electrical charge; when connected with a *resistor* it can be used as a *time delay* in an *electronic system* 30–33

centripetal force: the force that provides the *acceleration* needed to keep an object moving in a circular path; the force is directed towards the centre of the circle 53–57

centre of mass: the point through which the weight of an object can be taken to act, so far as its turning effect or *moment* is concerned 46–47, 50–51

collisions: see *elastic collision* and *inelastic collision*

colour code: a system of coloured stripes that is used to indicate the resistance of a fixed *resistor* 25

continental drift: the idea, proposed by *Wegener*, that land masses on the Earth are slowly moving 75–77

convection currents: the movement of a substance that can flow (e.g. the upper part of the Earth's *mantle*) caused by differences in temperature between its parts 70–71, 76

converging lens: a lens that causes rays of light that were parallel before they reached the lens to converge to a point (*focus*) after passing through it 36–43

crust: the outer part of the Earth, made from solid rock 64–65

D

diode: a device that will allow an electric current to flow through it in one direction only; can be used to protect a *transistor* when this is used with a *relay* 29

diverging lens: a lens that causes rays of light that were parallel before they reached the lens to diverge as if they came from a point (*focus*) after passing through it 36–37, 39

E

earthquakes: sudden movements of the Earth's *crust* 68, 72–73

elastic collision: a collision in which the total amounts both of *momentum* and of kinetic energy remain unchanged 62–63

electronic system: a system, comprising *input sensor(s)*, a *processor* and an *output device*, designed to respond in a particular way to change(s) in the surroundings 10–35

eye: a biological optical device that uses a *converging lens* to produce a *real image* 43

F

focal length: the distance between the centre of a lens and its *focus* 40

focus: the point to which light rays that were originally parallel are made to converge after passing through a *converging lens* (or from which they appear to diverge after passing through a *diverging lens*) 37

I

image: see *real image* and *virtual image*

inelastic collision: a collision in which the total amount of *momentum* remains unchanged but the total amount of kinetic energy is reduced 62–63

input sensor: the part of an *electronic system* that detects a change in the surroundings and indicates this by changing the input it provides to a *processor* 10, 12–13, 22, 27

L

LDR: short for *light dependent resistor*

LED: short for *light emitting diode*

lever: a device for increasing, or decreasing, the *moment* of an applied force 48–49

light dependent resistor: a device whose resistance decreases as the intensity of the light falling on it increases; it can be used as one of the *resistors* in a *potential divider* to form a *light sensor* circuit 10, 24

light emitting diode: a *diode* that emits light when a current passes through it; it can be used as an *output device* 12, 24

light sensor: an *input sensor* that is designed to detect changes in illumination (light level); normally made using a *light dependent resistor* as one of the resistors in a *potential divider* 24, 28–29

lithosphere: the Earth's *crust*, and the part of the *mantle* immediately below, that make up the mobile *tectonic plates* which form the outer part of the Earth 68

logic gate: an electronic device (e.g. an *AND gate*, an *OR gate* or a *NOT gate*) whose output depends on its input(s); they are used in the *processor* circuits of *electronic systems* 14–19

M

magma: molten rock from below the Earth's *crust* that reaches the surface when a *volcano* erupts and through *sea floor spreading* 65, 71, 75, 77

magnetic switch: a switch that closes when in a magnetic field; can be used as an *input sensor* (to detect the position of e.g. a door) in an *electronic system* 13, 22

magnification: the factor by which the size of an object must be multiplied to give the size of its *image* 40

magnifying glass: a *converging lens* used to produce a magnified, *virtual image* of an object 42–43

mantle: the layer of rock between the Earth's *crust* and core; though it is solid, it can flow slowly and becomes liquid *magma* when the pressure on it is removed 64, 70, 77

moisture sensor: a device whose resistance decreases when it is damp; it can be used as one of the resistors in a *potential divider* to form an *input sensor* circuit 26

moment: the turning effect of a force: the size of a moment (in newton metres, N m) is the size of the force (in newtons, N) multiplied by the perpendicular distance (in metres, m) between the line of action of the force and the pivot point 45, 48–50

momentum: the mass of an object multiplied by its velocity; during elastic collisions, inelastic collisions and explosions the total momentum of the objects concerned does not change 58–63

N

NOT gate: a *logic gate* whose output is on/high/1 when its input is off/low/zero, and vice versa; can be used in the *processor* circuit of an *electronic system* 15, 18

O

OR gate: a *logic gate* whose output is on/high/1 when input A <u>or</u> input B (or both) are on/high/1; can be used in the *processor* circuit of an *electronic system* 15, 18

output device: a device that is controlled by the *processor* of an *electronic system* and which acts on the surroundings when it is switched on 10, 12–13

P

plane of symmetry: a flat plane dividing a solid (3-dimensional) object into two mirror-image halves; the *centre of mass* of the object lies on this plane 47

potential divider: a circuit in which a voltage is shared by two components in proportion to their resistance; often used in *input sensor* circuits 23–27

processor: the part of an *electronic system* that receives signals from *input sensors* and 'decides' whether *output devices* should be switched on or off; may be made up of *logic gates* or be a *transistor* 10, 14–15

R

real image: an image that is formed (e.g. by a *converging lens*) when rays of light coming from a point on an object actually meet; an image that can be formed on a screen 38–41

relay: a device that uses the small current from the *processor* of an *electronic system* to switch on a larger current required by an *output device* 20–21, 28, 29

resistor: a component used to reduce the current through a circuit or, when connected in series with another resistor, to form a *potential divider* 7, 23, 25

S

sea floor spreading: what happens below the oceans when *tectonic plates* move apart; *magma* moves up into the space to form new rock 75, 77

sensor: see input sensor

stability: the more stable an object is, the more difficult it is to make it fall over; the wider the base of an object, and the lower its *centre of mass*, the greater its stability 50–51

subduction: the movement of oceanic *crust* below continental crust at places where *tectonic plates* are moving together 76

T

tectonic plates: the separate pieces of the Earth's *lithosphere* that are slowly moving all the time 68–72, 76–77

temperature sensor: an *input sensor* that is designed to detect changes in temperature; normally made using a *thermistor* as one arm of a *potential divider* 11, 25, 27

thermistor: a device whose resistance decreases as its temperature increases; can be used as one of the resistors in a *potential divider* to form a *temperature sensor* circuit 10, 25

thermostat: a mechanical device or an *electronic system* designed to keep a temperature at a constant level 8–9, 11

tilt switch: a switch that is open or closed depending on its orientation; can be used as an *input sensor* in an *electronic system* 13, 22

time delay: a circuit, often using a *capacitor* and a *resistor*, that is added to an *electronic system* to produce a delay between an *input sensor* registering a change and an *output device* being switched on (or off) 30–33

transistor: a component that allows a current to flow through it when the input to its base is on/high/1 and so can be used as an electronic switch, i.e. as a *processor* in an *electronic system* 28

truth table: a table that shows (using 1 = high/on and 0 = low/off) how the output of a logic gate (or combination of *logic gates*) is related to the various combinations of inputs 18–19

V

velocity: the speed of an object in a particular direction 52–53, 58

virtual image: an image that is formed (e.g. by a *diverging lens* or a *converging lens* used as a *magnifying glass*) when rays of light coming from a point on an object appear to come from a different point; an image that cannot be formed on a screen 39

volcano: a place on the Earth's *crust* where *magma* rises up from the *mantle*, breaks through the surface (erupts) and cools to form a cone-shaped mountain 65–67

W

Wegener: the scientist who first proposed the idea of *continental drift* as a serious scientific theory, although it wasn't generally accepted until *sea floor spreading* was discovered about 50 years later and explained in terms of moving *tectonic plates* 75